The Modern Garden

The Modern Garden

The Outdoor Architecture of Mid-Century America

Pierluigi Serraino

New York · Paris · London · Milan

First published in the United States of America in 2024 by
RIZZOLI INTERNATIONAL PUBLICATIONS, INC.
300 Park Avenue South, New York, NY 10010
www.rizzoliusa.com

Photography and drawings are credited on p. 222

Publisher: Charles Miers
Editor: Douglas Curran
Production Manager: Kaija Markoe
Managing Editor: Lynn Scrabis
Proofreader: Sarah Stump
Design Manager: Tim Biddick

Designed by Brett MacFadden, MacFadden & Thorpe

Printed and bound in China

2024 2025 2026 2027 2028 / 10 9 8 7 6 5 4 3 2 1

ISBN-13: 978-0-8478-3588-1
Library of Congress Control Number: 2024933832

Front Cover image:
Photograph by Ezra Stoller (see p. 141)

Half-title Page image:
Ford Foundation, New York, 1967
Architect: Roche & Dinkeloo
Landscape Architect: Dan Kiley
Photographer: Ezra Stoller

Title Page image:
Chinn Garden, San Francisco, 1950
Architect: Unknown
Landscape Architect: Robert N. Royston
Photographer: Ernest Braun

Back Cover Image:
Hall Residence, Los Angeles, 1962
Architect: Edla Muir
Landscape Architect: Edward Huntsman-Trout
Photographer: Maynard Parker

CONTENTS

Preface

ARCHITECTURAL PHOTOGRAPHY ARCHIVES have been a rich source of both content and reflections for me. As I continue to grapple with the far-reaching consequences of the photographic medium on our collective environmental awareness, new realizations surface following years of ruminations on what I have seen and keep finding out in these repositories of memory. This book is an offshoot of that extended processing in my unconscious.

A short anecdote will explain what triggered this publication. Being familiar with the photography of Morley Baer, I had seen the arresting images he took of the Alumni House at the University of California at Berkeley, a project designed by architect Clarence Mayhew and completed in 1956. That representation was all about the architecture showcasing all its exterior and interior elegance, with virtually no attention to the surrounding landscape. At a later time, while doing research in the archives of architectural photographer Rondal Partridge in Berkeley, I stumbled onto the pictures of a garden designed by H. Leland Vaughan for the Alumni House. In Partridge's images, I could barely see the building, as the focus was the garden. The architecture, whose identity was recognizable to me only because I had seen the structure in person, was the faded background of the photographic narration of the landscape.

Two highly esteemed photographers illustrated the same space, with each negating the existence of the other's evidence in being part of one and the same environment. It was either about the architecture or about the landscape. Not both. How could it be possible? What were the long-term ramifications for my understanding of space and for my priorities as an architect, as well as those of clients and even others, in attending to the complexities of the public realm?

This book is an invitation to see and appreciate space in its entirety, beyond disciplinary strictures, and to escape the conditioning of the mechanical medium so as to reengage us in the physical world.

Hyatt Regency, San Francisco, CA, 1972
Architect: John Portman
Landscape Architect: Unknown
Photographer: Unknown

PART 1

Garden in the Camera: Photography and Landscape

The beauty of our land is a natural resource. Its preservation is linked to the inner prosperity of the human spirit.[1]

LYNDON JOHNSON

PREVIOUS SPREAD:
Location Unknown, 1965
Architect: Unknown
Landscape Architect: Unknown
Photographer: Ezra Stoller

OPPOSITE:
Lillian B. Ladd House, Pasadena, CA, 1953
Architect: Thornton Ladd
Landscape Architect: Thorton Ladd
Photographer: Ezra Stoller

THE INTRODUCTION OF architectural photography has indirectly and inadvertently transitioned humankind from a culture of space to a visual culture of space. That means that the learning of a building, a site, a natural landscape, and everything in between through the camera largely anticipates its direct experience. Most likely, this was a side effect of the original intent of this ingenious invention, but a portentous one nonetheless, and of remarkable historical consequences.

The first casualty of this epochal shift in hierarchical perception was landscape architecture.

Before photography, anyone wanting to learn about destinations mentioned in books or through oral accounting came to know of any of these places through direct experience. Going there in person made the environmental dimension of space across all its scales, inclusive of architecture, patently obvious to the observer. Renaissance master Andrea Palladio got his education in Roman architecture and its urban character by visiting the Eternal City and measuring its classical monuments himself to capture its proportions, record those ratios in his notebooks, and develop an environmental sense that would eventually inform his designs. Through that immersive encounter, he understood how those buildings were sitting on the land. He developed a bodily knowledge of those achievements and internalized their traits. Many others did the same before and after him in a similar vein, even when lesser results ensued in their respective creative endeavors. The Grand Tour served an analogous purpose, exposing the members of the nascent upper class of seventeenth-century European society and beyond to the landmarks of what the canon was at the time.

Since photography entered the public realm in the nineteenth century, however, knowledge of distant places and faraway architecture opened up new vistas in both scholarly and popular imagination from the comfort of their stationary position. As this new technology emerged, the awareness of "otherness," of the exotic, of places beyond the routine circuits in the prevailing travel customs of a humanity in rapid transformations became a fixture of the information society at its infancy. The limits of the medium employed decisively defined its perceptual boundaries for the viewer, patterning its visual content in very specific, and often reductive, ways. Only a subset of the vastness of the places represented fell within its viewpoint. Furthermore, what was in the camera was also dependent on what the photographer deemed crucial, assigning a secondary position to other content.

In the main, the photographer places the camera with the vision plane parallel to the front of the buildings to be depicted. In that configuration between photographer and object to be represented,

**Gade House, Emerald Bay,
Laguna Beach, CA, 1962**
Architect: Ladd & Kelsey
Landscape Architect: Thornton Ladd
Photographer: Ezra Stoller

the Z axis plays a dominant role in organizing what is portrayed and how it is seen in a picture. A side effect, and hardly an insignificant one, is that within the photographic frame, the representation of the ground plane is but a small percentage of the visible content of the image. The contours of the architecture represented in the viewfinder approach the edges of that artificial boundary that photography generates, determining a perceptual tension that in reality is nowhere to be found in an actual visit to those same places the medium memorializes. Through photography, the site has progressively become inconspicuous, recessed in the background of collective awareness in favor of the aggrandizing of the building. The garden is routinely at the edges of the photographic frame, in a peripheral position from where the viewer focuses first, that is, the center of the photograph where architecture resides.

Experientially, however, the opposite is true. In fact, the perceptual ratio of building versus setting is typically reversed, unless the structure is located in dense urban settlements. All that is contained in the peripheral field of vision, reduced to quasi-invisibility in the picture, reappears upon site visits. A case in point is the widely acclaimed Irwin Miller House (1953–57) in Columbus, Indiana, by Eero Saarinen, with the interiors by Alexander Girard. While the architecture of the residence is fixed in the historical canon through the definitive images of architectural photographer Ezra Stoller, the sheer dimension and complexity of its site, designed by Dan Kiley, far bigger than the actual size of the house, escape environmental understanding through those pictures in the absence of an in-person visit.

The perceptual displacement of photography thus grafts an artificial order on space reading foreign to the space formation of the natural site. Landscape lacks objecthood and is innately relational: it connects by default. Landscape architecture is a relational form of design. It implies multiplicity of specific natural and human-made spatial events integrated in the totality of the environment. Nature comes with no predetermined hierarchy, whereas photography imposes one, being an inherent part of its graphic language. The garden, however, as bounded landscape sets the terrain for another design order within the legal boundaries of the site and the hard edges of the building. In landscape design, there is no front or back, but only the ground plane amenable to innumerable spatial permutations. Gardens are modulated through plant associations that create texture, as well as the imaginative negotiation between hardscape and landscape. While, time and again, architects have used the golden section, the Fibonacci series, and similar mathematical a priori factors on organization of the plans and facades' layouts, landscape architects appear free from these preconceptions. Within this process, photography tends to promote the sharp division of perception between building and garden, rarely engaging both and giving them the same weight.

Charles Beber House, Coconut Grove, FL, 1967
Architect: Peter Jefferson
Landscape Architect: Unknown
Photographer: Ezra Stoller

As architects generally become aware of present and past projects through photography, this basic fact has momentous ramifications in the production of new works. As opposed to its spatial attributes, the graphic nature of architecture—primarily its exterior envelope—takes over in the hierarchy of formal concerns of the contemporary practicing designer. The prevailing tendency is to conceive of architecture as a canvas on which to display a wide array of random patterns of fenestrations, prefabricated panels, and associated building components, at the expense of an environmental as well as spatial reading of the site of which architecture is a subset. Such generational loss of understanding of the tight connection between architecture and landscape unfolded in temporal increments lasting almost two centuries. The generational aggregate of this myopic view of

Julimar Farm Gardens, Queens, NY, 1964
Architect: Unknown
Landscape Architect: Edward Durrell Stone Jr.
Photographer: Ezra Stoller

design is generative of reductive solutions on each side. At present, the cumulative layering of this misrepresentation, a societal shift of focus through the technology of mechanized vision from the land to the building, comes with a significant social, cultural, and environmental cost. Historically it has profoundly impacted the organization of design practices entrenched in the limits that professionalism established with the rise of industrialization and mass society.

Architects overestimate the value of the treatment of the enclosure, and landscape architects stake out a reduced footprint for their own agency in the design process, like a separated couple under the same roof but sharing subsets of common spaces. This disconnect delivers fragments of isolated universes, the architectural and the landscape ones, missing a structural link to their

Wellington Henderson Residence, Hillsborough, CA, 1957–60
Architect: Wurster, Bernardi & Emmons
Landscape Architect: Thomas D. Church
Photographer: Morley Baer

Macht House, Baltimore, MD, 1957
Architect: James Rose
Landscape Architect: James Rose
Photographer: Ezra Stoller

adjacent areas of intervention. The togetherness of architecture and landscape, the indissoluble environmental unit that we as individuals experience consciously and unconsciously every day, is routinely lost in the disciplinary representation of each field. Architects draw their structures with a topographic and infrastructural sense, but in the main less attuned to the landscape sense, that is the constant rebalance of hardscape and landscape at the service of space.

Although landscape architecture is the interface between the building and the site, its appreciation remains elusive largely through this aforementioned process. The adoption of photography as the dominant agent of space representation bestows an inordinate amount of attention to the architectural object as a self-referential sculpture, at the expense of its integration in the environment of which the architecture is part. Statistically speaking, the visual representation of the built environment, both in drawings and photographs, thus overwhelmingly places asymmetrical focus on the building, assigning a secondary place to its surroundings in matter of space formation. Even though one cannot exist without the other, this pattern of perception is pervasive and academically established. A few examples exemplify the point.

The Villa Rotonda in Vicenza, by Andrea Palladio, has been widely admired since the late sixteenth century for the numerical beauty of its pure geometry; it also has maximum impact on the visitors because it is located at the summit of a terrain, creating a powerful sense of anticipation through the ascent from the access at the bottom of the precinct toward this masterwork. That topographic condition, designed as much as the villa was, is rarely, if ever, mentioned in the aesthetic appreciation of this landmark, but it is undeniably there. The photographs of that architecture rarely account for that commanding terrain configuration and building placement. The Rotonda would be a significantly lesser statement on a different site. Their inseparable togetherness is at the basis of its powerful appeal across time.

Similarly, Villa Mairea, by master architect Alvar Aalto, is inlaid in the arresting setting of the Finnish landscape it is so very much part of. Its location in Noormarkku, Western Finland, makes it challenging for a large segment of the population to visit this masterpiece. As it is located outside more common circuits, planning and hours of driving are required to get there and see it. As a result, very few of the Modern masters as well as Aalto's contemporary historians have ever visited that house, included in virtually every architectural history book since its completion in 1939. Hence the reliance on photographic images for its assessment. Yet Aalto put landscape at the center of his concerns in his conceptual framework on design. Even in the sheets representing the floor plans, elevations, and study sketches of that canonical building, the presence of the trees is carefully and consistently drawn. The vanishing point of most of

PREVIOUS SPREAD:
Everett Residence, Orinda, CA, 1950
Architect: William Russell Everett
Landscape Architect: William Russell Everett
Photographer: Rondal Patridge

OPPOSITE:
Wohlstetter House, Los Angeles, 1954
Architect: Josef Van der Kar
Landscape Architect: Garrett Eckbo
Photographer: Julius Shulman

OPPOSITE:
Carter House, Tulsa, OK, 1962
Architect: O'Neil Ford
Landscape Architect: Lambert Landscaping, Co.
Photographer: Ezra Stoller

ABOVE:
Hogel Art Studio and Garden, Woodside, CA, 1970
Architect: John Matthias
Landscape Architect: Unknown
Photographer: Glenn M. Christiansen

Deering Milliken Research Corporation, Spartanburg, SC, 1960
Architect: Skidmore, Owings & Merrill
Landscape Architect: Unknown
Photographer: Ezra Stoller

the photographs representing that project, however, remains the building, both then and now. The authoritative role that nature plays in the appreciation of the Villa Mairea, self-evident when visiting that piece of architecture in person, becomes remarkably inconspicuous through the images circulating in the world of publications and media. Between the property line of a parcel of land and the building edges, in fact, an entire spatial world exists, with its three-dimensional articulation, texture, scale, and narrative sequence. You see the forest before you see the building. The architectural experience has already started with the ascending path leading to the residence. Aalto matched that same sequence revealing the building amid the landscape in the Maison Louis Carré in Bazoches-sur-Guyonne, France, completed in 1959.

Hutton Residence, Montecito, CA, 1962
Architect: Lutah Maria Riggs
Landscape Architect: Wendell Richard Gilbert
Photographer: Julius Shulman

Garden, Mineola, NY, 1950
Architect: James Rose
Landscape Architect: James Rose
Photographer: Ezra Stoller

A lesser-known but remarkable piece of architecture is the Noyes Residence in New Canaan, Connecticut, by Eliot Noyes. A classic of the Mid-Century era in the Northeast, it is a textbook case of a design structured in plan around the commanding geometry of the square. Its layout is rigorous and functionally impeccable, with elevations whose mighty stone walls, textured with light and shadows on a clear day, cast an irresistible spell on its visitors. The house, once again, is magnificently sited on the ground plane, with a carefully planted ring of landscape buffering it from the street and its adjacent lot. From the interior patio, cutouts of nature visible from all sides are an integral feature of experiencing the space. Peter Rolland was the landscape architect who contributed to the choreography of that setting, but scant attention is given to his input in the historical credits for this landmark.

Silverstein House, Beverly Hills, CA, 1962
Architects: Jack Charney and Sanford Kent
Landscape Architect: Eijiro Nunokawa
Photographer: Ezra Stoller

All these cases share one common denominator: the zone between the building's edge and the street, so rich in optical stimuli and otherwise, is ordinarily perceived in a state of distraction or as secondary in spatial importance to the building. That is the territory of landscape architecture.

The invisibility of landscape architecture is less acute in the public realm. There, photographic documentation offers extensive records of piazzas and streets, showcasing their structural function as catalysts for collective contact. But even then, these voids are routinely undetected in the admiration for the structures. Two examples suffice to demonstrate this point. The archetypal strength of the Casa del Fascio in Como, by the Italian rationalist Giuseppe Terragni, is amplified through the piazza in front of it, approximately matching in depth that of the building. Similarly, the iconicity of the Pompidou Center in Paris, by Renzo Piano, would be significantly lessened in the absence of the slanted plane—the "piazza," as the recently deceased architect Richard Rogers used to call it—leading from Rue Saint-Martin toward the museum entry at the lower level. This widespread pattern is reflected in the credits of those who author the project. While many know the Mid-Century landmarks of Eero Saarinen, Skidmore, Owings & Merrill, and Edward Durell Stone, only a few also know the landscape architects who did the master plan for all the outdoor areas of those projects. The consequences of this oversight are patently obvious in the bulk of construction globally, where concerns for the outdoors is an afterthought when compared to the center stage that the actual building has as a real estate venture. Landscape is deemed an amenity rather than structural to the design of space as place.

The fact that landscape architects and architects hire the same photographers is of remarkable significance. Thomas Church would hire Maynard Parker and Rondal Partridge, who respectively photographed the exemplary works of Harwell Hamilton Harris and John Carl Warnecke. Garrett Eckbo would employ Julius Shulman, who provided the iconography of California Modernism. Lawrence Halprin would recruit Morley Baer and Ernest Braun, who gave the definitive images of Sea Ranch and the look of *Sunset* magazine. James Rose would use the mastery of Ezra Stoller, who memorialized American modernity through his camera, to record his gardens in pictures. These photographers were all operating within a generation and inventing a way of looking at Modern space, irrespective of whether it was architecture or landscape. In the vast majority of the exposures they conceived to frame landscape architecture, their camera is tilted. Seeking high points, terraces, and any designed or circumstantial change of elevations that would afford a

vantage point for the camera to reveal the otherwise invisible world of landscape architecture linked those vastly different photographic signatures. Just as with buildings, these image makers stylize landscapes through photography.

The personal preferences of each photographer shape the construction of the image, whether the subject is architectural or landscape. Maynard Parker was one of the oldest architectural photographers of his generation. Much of his work was for *Architectural Digest*, an interiors magazine. Focus on decoration was an integral component in his photography. His landscape pictures cater toward taste formation. Rondal Partridge was the son of Imogen Cunningham. Fine arts as a primary intent far preceded the need for income that architectural photography provided; the desire to produce an artistic statement was

ABOVE:
Kookish Burnell Garden, Bel Air, CA, 1952
Architect: Samuel A. Marx
Landscape Architect: Garrett Eckbo
Photographer: Julius Shulman

OPPOSITE:
Hester House, San Diego, CA, 1962
Architect: Henry Hester
Landscape Architect: Harriett Barnhart Wimmer
Photographer: Julius Shulman

Taylor House, Location Unknown, 1960
Architect: Cliff May
Landscape Architect: Unknown
Photographer: Julius Shulman

consistently present in his compositions. Julius Shulman delivered a hedonistic rendition of modernity, where lightness, consumption, and joy imbued so many of his classic photographs. Morley Baer and Ernest Braun were at heart nature photographers; nature featured prominently in most of their images. Ezra Stoller, who developed his uncanny ability to capture Modern space with extraordinary consistency throughout his career, offered the still contemplation of space as the hallmark of his photographic signature. With all their pronounced differences, these image makers were supplying images to the same magazines and publications. Curiously, it is in a curated selection of their work that the loss of unity of architecture and landscape can be rediscovered and made evident.

Bick Residence, Brentwood, CA, 1951
Architect: John Lautner
Landscape Architect: Unknown
Photographer: Julius Shulman

In truth, a great many architectural photographers tend to create visual journeys of the sites they will take pictures of. They map out mentally or materially on a sketch plan a sequence of meaningful viewpoints, and they carry out that route during their assignment. Photographers act as the first curator in structuring the visual consumption of that space for the general public. Within the assortment of pictures provided to publications, editors make a further selection, picking those images with maximum graphic impact for publication. The result is a few stills of built reality that serve editorial purposes specific to the readership the magazines and publications cater toward, at the expense of apprehending the integrity of the site along the way. Photographers, however, have a clear sense of the tight link between buildings and their sites, and within the confines of the medium they attempt to deliver that message to the viewers.

Whoever hires the photographer also affects the representation of what is being photographed. If the architect initiates that relationship, emphasis is on the building, whereas the opposite occurs when the landscape architect does. While in the economics of space representation this imposed focus is understandable, it is in the consumption of these images that aspects of the disconnect between architecture and site are most apparent. Statistically, buildings are more known, and therefore acclaimed, than their landscape. Although uncommon, however, some outdoor areas have become better known than the architecture itself. A case in point is the Donnell Garden in Sonoma, designed by Thomas Church with a young Lawrence Halprin. The garden is far more celebrated than the architecture of the residence by George Rockrise, a reputable architect of the Mid-Century period. In a

ABOVE:
Case Study House No. 28, Thousand Oaks, CA, 1966
Architect: Buff & Hensman
Landscape Architect: Jack W. Buktenica
Photographer: Julius Shulman

OPPOSITE:
Perkins House, Pasadena, CA, 1955
Architect: Richard Neutra
Landscape Architect: Richard Neutra
Photographer: Julius Shulman

McIntyre Residence, Hillsborough, CA, 1960
Architect: Joseph Esherick
Landscape Architect: Lawrence Halprin
Photographer: Ezra Stoller

preceding similar case, the design of the garden for the Ralph K. Davies House in Woodside, California, was completed by Thomas D. Church before Anshen & Allen designed the first project that launched their career. Museum catalogs contribute as well to this chronic divide between architecture and landscape. The McIntyre Garden in Hillsborough, California, completed in 1960, is part of the McIntyre Residence, designed by noted architect Joseph Esherick. Yet there is no mention of the architect in that title.[2]

The invention of Modern architecture called for a whole new world of forms. Everything had to be redesigned to be part of a coherent whole—interior furniture and gardens included. Axial organization of outdoor spaces imbued with the historical spirit of preindustrial centuries were poor allies in the spatial revolution that occurred in the design world at the dawn of the twentieth century. Landscape architecture inevitably engages the organic and the mingling lines of nature. This formal repertoire clashed with the recasting of space on a purely industrial basis. The assertive abstractions of the avant-garde modernists were at odds with the lighting effects and the shadow play of foliage and texture. And yet, there are no landscape architects in the Bauhaus legacy. Discourse on the organic, a foundational aspect of landscape architecture, is absent from a culture where mass production and the machine has center stage. The garden, as a figure of speech, became invisible in the Modern movement. But the open plan in architecture needed an outdoor design expression as well. Its correlate was the open garden. Their appearance in architectural history occurred at the same time. Modern painting provided formal references for the construction of new imagery. In replacing the axial organization of the outdoors from previous centuries, a novel repertoire of relationships and forms emerged, rooted in the visual explorations of fields of colors, floating geometries, and free form. Figuration over issues of resource management of the land was the driver of the organization of layouts that were captivating in their relational novelty.

Mies van der Rohe provided a powerful rendition of the architecture of hardscape in the Barcelona Pavilion at the 1929 International Exposition in that Spanish city. He established the modernist palimpsest for the attitude of the avant-garde toward the treatment of open spaces. Uncluttering became the new creed, and many followers diligently walked that path. Photography has a major role in the dissemination of this attitude, since the pavilion was dismantled shortly after the exposition ended. The Modern garden inherited part of this totalizing attitude. It purged any trace of the romantic landscape to take on the signs from various sources. Abstract painting was an inspiration to some of the most radical landscape architects. A novel world of formal relationships ensued, intentionally detached from any

Donnell Garden, Sonoma, CA, 1947–54
Architect: George Rockrise
Landscape Architect: Thomas D. Church
Lawrence Halprin, assistant

Donnell Garden, Sonoma, CA, 1947–54
Architect: George Rockrise
Landscape Architect: Thomas D. Church
Lawrence Halprin, assistant
Photographer: Rondal Partridge

Frizani Residence, Hollywood, CA, 1954
Architect: John Kewell
Landscape Architect: Al Heineman
Photographer: Julius Shulman

Beaux-Arts plan. Absent, however, were concerns about ecology and the vocabulary associated with those preoccupations. Words like habitat were largely missing from the modernist discourse on landscape.

The management of the site changed in the late 1930s with the dawn of architectural modernity. A few members of the generation of landscape architects that entered the field in the thirties were particularly articulate in making arguments for a new sensibility in this area. Garrett Eckbo wrote: "First, what do we mean by 'landscape architecture'? In the broadest definition the landscape is everything one can see from a given station point or series of them, and thus is compounded by building and landscape development."[3] He later added: "The essence of the space concept is the

Frizani Residence, Hollywood, CA, 1954
Architect: John Kewell
Landscape Architect: Al Heineman
Photographer: Julius Shulman

consideration of the voids or air spaces in the design as positive elements, to be shaped for their own sakes, rather than as negative by-products of the arrangement of solids."[4] This newfound awareness will be the source of an astonishing chapter of space making that cemented the enduring legacy of postwar American architecture. Some architects never lost sight of the tight connection between architecture and nature. Aalto wrote: "The garden (or courtyard) belongs to our home just as much as any of the rooms."[5] Occasionally, structural engineers also realize the landscape import of bridges in their design. Noted prestress expert T. Y. Lin had the insight to recruit Lawrence Halprin to site the 23rd Avenue Bridge in Oakland, California. As rare as these examples are, they are instances of integrated design imbued with environmental awareness.

OPPOSITE:
Hook Guest House, Siesta Key, FL, 1952–53
Architect: Paul Rudolph
Landscape Architect: Paul Rudolph
Photographer: Ezra Stoller

ABOVE:
Yew House, Los Angeles, CA, 1957
Architect: Richard Neutra
Landscape Architect: Richard Neutra
Photographer: Julius Shulman

Edward Thrower House,
Location Unknown, 1961
Architect: Unknown
Landscape Architect: Unknown
Photographer: Ezra Stoller

The architectural drawings of the floor plans of Frank Lloyd Wright, Alvar Aalto, and Richard Neutra routinely feature the landscape treatment. Contrary to the standard practice of landscape architects to block out the design content of the architectural interiors, excluding the building enclosure, and of architects doing exactly the same in their own drawings, these three Modern masters bridge the reality of architecture and its sight into one document restoring the integrity of the environmental experience. Similarly, the drawing plan of Manitoga, the estate of industrial designer Russel Wright and Mary Small Einstein Wright, prominently features the existing topography and the garden. This total work of art executed between 1941 and 1961 reflects a unified vision and coherence, connecting landscape, architecture, interior, and product design.

Oakland Museum of California, Oakland, CA, 1968
Architect: Roche & Dinkeloo
Landscape Architect: Dan Kiley
Photographer: Ezra Stoller

Henry McIntyre Residence, Atherton, CA, 1968–70
Architect: Dimitry Vedensky
Landscape Architect: Thomas D. Church
Photographer: Rondal Partridge

In the postwar period, there were sparse hints that landscape architecture was an art form with its own standing within the design universe. Frank Harris and Weston Bonenberger's *A Guide to Contemporary Architecture in Southern California*, published in 1951, featured separate sections on landscape architecture sites to visit, reversing the editorial hierarchy by crediting the landscape architect first and the architect second. It was a rare instance of both acknowledging the integration of architecture and landscape as a seamless spatial continuum and the realization that modern space, as opposed to divisions among the design professions, was the driver. But it was even rarer to find a landscape architect authoring articles published in architectural magazines. *Pencil Points* was among the first periodicals to give editorial space to James Rose, Garrett Eckbo, and others to write about various aspects of the interface between architecture and landscape. Similarly, it was uncommon to find an architect publishing essays

OPPOSITE:
Ewing House, Coconut Grove, FL, 1957
Architect: Alfred Browning Parker
Landscape Architect: Unknown
Photographer: Ezra Stoller

ABOVE:
Schutt House, Los Angeles, 1946
Architect: Burton Schutt
Landscape Architect: Unknown
Photographer: Julius Shulman

in landscape architecture journals. This lack of dialogue between these two camps is entrenched and hard to eradicate. It is therefore particularly noteworthy that Eckbo wrote the foreword to the only monograph on the architect Joseph Allen Stein, aptly titled *Building in the Garden*.

The Modern Garden squarely tackles such historical editorial imbalance by recalibrating the visual delivery of settings that became foundational for the construction of a design culture in North America at mid-century, and later absorbed worldwide. In bringing forward the organic and hardscape context of which both notable and lesser-known architectural landmarks are part, this publication offers a more balanced understanding of what is at the roots of their longevity in design history. The 166 designs of mostly residential projects presented here and built during the Mid-Century Modern period across the United States, along with one example in Puerto Rico and one in Turkey, highlight their structural integration of the land and the architecture it hosts on its ground, thus reconfiguring the pattern of appreciation of these milestones in the design field.

ABOVE:
Weyerhaeuser Corporate Headquarters, Tacoma, WA, 1971
Architect: Skidmore, Owings & Merrill
Landscape Architect: Sasaki, Walker and Associates (SWA)
Photographer: Ezra Stoller

OPPOSITE:
Levitt Office Building, Lake Success, NY, 1965
Architect: Edward Durrell Stone
Landscape Architect: Edward Durrell Stone Jr.
Photographer: Ezra Stoller

Through the proposed extensive appraisal of the outdoor architecture in its numerous iterations in North America, this editorial offering features content accessible at multiple levels. It is intended to be both an effective source of knowledge and a visual delight for consumers and producers of design. It showcases a curated collection of exemplary works, whose influence on the subsequent development of environmental design is demonstrated in the shift of the collective focus from the building to the setting, as it occurred in the early 1960s. It aims at promoting a culture of integration of the various design areas, casualties up to now of the specialization that has so deeply affected architecture as an art form at the expense of its overall spatial character. It acknowledges the central role that the outdoors has in bettering the qualitative nature of the human experience.

After personal visits to Villa Rotonda, Villa Mairea, the Casa del Fascio, and the Pompidou Center, with all their profoundly marked differences, the common experience I had was the revelation of the settings these structures were embedded in. I was simply unaware of what these masterpieces were part of until I saw them, even though I knew of them through photography. In reference to residential architecture in particular, the bulk of the images featuring outdoor areas trigger a desire in viewers to become part of the scene, to a much greater degree than even residential interiors of the same architectural projects. Unconventional silhouettes of pools, lush gardens, original material transitions through abstract patterns, unusual open-air stairs, and the placement of design elements in harmony with the surrounding natural environment are sources of positive associations, where leisure, personal memory, and the realization of an idealized life energize viewers.

A few personal remarks are in order. Decades of site visits to see buildings in person cemented my realization that the surrounding environment is as important as the piece of architecture itself. Could Fallingwater, by Frank Lloyd Wright, be as impactful as it is if there were radically different surroundings? Being an avid book collector and an author of architectural books, I absorb architectural information on an ongoing basis. Why, then, am I surprised virtually every time I visit these milestones of architectural history? Why am I unaware until I see it in person that the Lovell Beach House in Newport Beach, California, by Rudolph Schindler, is in fact on a side street, missing the wide-open areas that definitive eye stoppers of the era and later times suggest? On a regular basis, as a practicing architect, I confront the challenges of making the case with clients to assign an adequate budget for landscape architecture. The customary attitude is that outdoor areas are an embellishment, a sheer afterthought once the structure has been set—in a nutshell, a nicety that is a

THIS PAGE:
Mariners Medical Arts Center, Newport Beach, CA, 1964
Architect: Richard Neutra
Landscape Architect: Unknown
Photographer: Julius Shulman

OPPOSITE:
Jewell House, Coconut Grove, FL, 1957
Architect: Alfred Browning Parker
Landscape Architect: Unknown
Photographer: Ezra Stoller

BELOW:
Dr. Charles Beber Porch, Coconut Grove, FL, 1965
Architect: Peter Jefferson
Landscape Architect: Unknown
Photographer: Ezra Stoller

OPPOSITE:
David Frame House, Houston, TX, 1961
Architect: David Frame
Landscape Architect: Unknown
Photographer: Ezra Stoller

Gilbert Hahn Jr. Residence,
Location Unknown, 1963
Architect: Unknown
Landscape Architect: Unknown
Photographer: Ezra Stoller

candidate for elimination in the next round of value engineering. As desirable as it might be, landscape design remains an amenity that can be done without, due to its cost. This is a prevailing stance, highly problematic for its fallout on the built environment. But even as a student in several architectural schools, I was never required to take a course in landscape architecture. From an educational standpoint, landscape was an irrelevant factor on my horizon as a committed designer.

Architectural photographers feeding the visual culture of our time deliver countless images of glazing systems, shadow boxes, rooftops, and skylines, statistically neglecting the representation of the ground plane where the occupants spend most of their time. When you visit architecture over decades, you develop a bodily knowledge of space navigation and scale of one's body in relation to the physical world. That accrued bodily knowledge becomes the basis of architectural appreciation. This, in my belief, is at the root of the placeness of much of current design production: a complex mix of concomitant factors relegating what is obvious to the background of our awareness.

Landscape architecture is a less favorable subject for the spectacle of contemporary media glamour. Its limited elevation curtails the hypes of the manufactured horizon of a reverse environmental hierarchy: building first, site a distant second, though the reverse is experientially obvious. The protagonist of the collective experience is the ground plane. Our feet are on the ground even when our eyes wander. The world beyond the architectural enclosure takes on a multiplicity of roles from the recreational to the contemplative, well beyond the strictly infrastructural function that the culture of modernity has assigned to the land. The recent pandemic has brought to the fore the vital importance of open spaces of any scale, both at an individual and at a community level. It has reinforced the crucial function of the necessity to bestow equal attention to what is outside as much to what is inside. Space is the common denominator of all that exists in the material world outside us. For centuries, architectural history has placed primary importance on the building, its patrons, and architects. Their context, that is, the fact that they exist as part of something bigger than themselves, has faded in the background. The pandemic, however, has made humanity aware of the importance of space, outdoor as well as indoor. Deflecting the attention on one to favor another comes at a price increasingly impossible to afford, culturally and ecologically in equal measure.

Riley Streate House, Miami, FL, 1952
Architect: Alfred Browning Parker
Landscape Architect: Unknown
Photographer: Ezra Stoller

Parker House, Miami, FL, 1952
Architect: Alfred Browning Parker
Landscape Architect: Unknown
Photographer: Ezra Stoller

Parker House, Miami, FL, 1952
Architect: Alfred Browning Parker
Landscape Architect: Unknown
Photographer: Ezra Stoller

Footnotes

1. Lyndon Johnson, *Beauty for America: Proceedings of the White House Conference on Natural Beauty*, Washington, DC, May 24–25, 1965 (Washington, DC: US Government Printing Office, 1965), 15.
2. Elizabeth B. Kassler, *Modern Gardens and the Landscape* (1964; rev. ed., New York: Museum of Modern Art, 1984), 24.
3. Garrett Eckbo, "What Do We Mean by Modern Landscape Architecture?," *Journal, Royal Architectural Institute of Canada* 27, no. 8 (August 1950): 268–69.
4. Ibid., 269.
5. Alvar Aalto, "From Doorstep to Living Room," in *Alvar Aalto in His Own Words*, ed. Göran Schildt (New York: Rizzoli, 1997), 51.

PART 2

Garden as Space: Outdoor Architecture

Gardens, like houses, are built of space. Gardens are fragments of space set aside by the planes of terraces, walls and disciplined foliage.[1]

JOSEPH HUDNUT

PREVIOUS SPREAD:
Guy Henle House, Cape Cod, MA, 1962
Architect: Unknown
Landscape Architect: Unknown
Photographer: Ezra Stoller

OPPOSITE:
George Wolff Residence, San Francisco, 1951
Architect: Erich Mendelsohn
Landscape Architect: Thomas D. Church
Photographer: Rondal Partridge

THE INTERCOURSE BETWEEN open and enclosed spaces has been a universal theme in architecture across times, technologies, and cultures. It is a permanent dialectical condition that any designer comes to grips with while in the process of shaping environments for human inhabitation. Even though they are two constants in the design of space, their relationship is historically variable to a remarkable degree. Emphasis of one over the other is the result of concomitant factors in a perpetual dynamic correlation with each other. Whether the spatial character of the garden grows out of the architectural expression of the building or vice versa, or they are born concurrently as part of a singular gesture, the resulting configuration reaffirms that architecture and landscape always come together. Their indivisibility is a first principle in the art of building.

As the spatial revolution affecting all domains of design was picking up speed in the early decades of the twentieth century, architectural discourse placed typology at the center of its debate, making it a dominant focus of inquiry at the expense of other areas of concern. The inward look at the architectural plan that the Modernists enacted then had its roots in the narrative structures of the early architectural treatises published in the Renaissance following the invention of the printing press. Leon Battista Alberti, Sebastiano Serlio, Andrea Palladio, Giacomo Barozzi da Vignola, Vincenzo Scamozzi—that is, the protagonists of the Italian Renaissance and Mannerism—followed this model. It was during the Age of Enlightenment, however, that the practice of representing architecture through stand-alone self-referential drawings, deprived of any information about their context, became the undisputed norm of the genre. Well-known are the matrices that the French architect and academician Jean-Nicolas-Louis Durand composed in an effort to recast the foundations of architectural design on rational tenets. Tables after tables of both theoretical layouts based on modular growth and historical examples of canonical buildings presented those designs as spatial axioms severed from their sites. Arrayed in rows and columns, plans and elevations of structures were illustrated in their orderly inevitability floating in the immaculate white space of the paper orphan of the physical surroundings they were an integral part of. Similar to then-nascent laboratory practices of preserving parts of, or whole, animal creatures and plants of the natural world in chemicals for scientific study, buildings were published in books as specimens detached from the larger portion of the world they belonged to. Through editorial artifice, canonical structures were turned into self-contained units forming a large repository of available instances demonstrating the rationality of universal design. Architects could mine this vast and methodically arranged resource so as to design their buildings shaped on this unchallenged model of excellence. Descriptive geometry provided

John Tuteur Jr. Residence, Napa, CA, 1967
Architect: Germano Milono
Landscape Architect: Unknown
Photographer: Rondal Partridge

the foundations for the reliable portrayal of architectural examples in their true dimension. The orthogonal projections of Gaspard Monge transformed architecture in Platonic solids levitating without gravity and with no ground plane to account for. In this orderly way of looking at the world, the formal irregularities of nature in all its material manifestations were out of character and therefore disposed of in architectural drawings. The building footprint was both the threshold and the containment of any architectural legitimacy, ousting the outdoor areas from the purview of the architect's attention. Architecture was in, and landscape was out.

As function rose to prominence in the preoccupations of the modernist architects, discourse on typology transitioned from the exact application of a formal a priori geometry to space optimization for best use given minimal means. However, the traditional approach of isolating architecture from landscape remained firmly in place. The primary goal was to maximize the utility of scarce square footage and to discover the functional constants historically determined through centuries of established use. Representatives of this prevailing line of research intentionally left out all concerns about the natural context that those building types were always integrated into as part of their own genealogical history. In their physical configuration, buildings and their sites were carrying together the heritage of their builders across time, delivering them to the following generations. But this environmental coupling was denied any rights of representation in architectural literature. This caused the aggrandizing of the building in the formal culture of space and the associated relegation of the land to opacity rather than an equal contributor in space making. Instead, identifying the most effective functional adjacencies internal to the individual plan components, applied in the same fashion to residential or nonresidential projects, was the litmus test for their validity as a coherent design proposition. In that inquiry, the garden was conceptually passed over as a meaningful area amenable to any lasting design intervention and was neglected altogether in architectural circles. In this peculiar view, the complex world of the exterior environment, starting from the perimeter of the building to the legal site boundaries, was assigned a secondary role in determining the quality of living, past and modern. The consequences of this collective choice were of historical import. At best, architects carried out a perfunctory reading of the site, barely the locus where to lay out the road for car access from the entry to the front door, a pervasive infatuation in the symbolic world of the modernists across the globe.

Metaphorically speaking, the outdoors was left out of the doors of architecture.

The separation of the building from the garden went hand in hand with the development of prototypes in architecture and industrial design. By default, a prototype is non-topical. Its existence builds on an internal logic—structural, functional, and economical—that provides its *raison d'être* as what it is while remaining fundamentally independent of context. Its placement in a location is indifferent to broader environmental considerations. As such, every setting works. Prototypes can be plopped into any topographic condition and located in every geographic coordinate, just like mass-produced chairs placed as needed whenever, wherever. A case in point was the birth of the prefabricated house, a turning point in demonstrating that residential design was in essence a construction problem rather than an architectural one. This newcomer in architectural discourse was expected to live up to inherent technical performances of building systems, uncommitted to any place, philosophically and materially. The building's orientation, the way it is sited on the terrain, spatial concerns on how it relates to what is outside of itself—all were purposefully left out from the problem areas. Its lack of situatedness was seen as a point of strength, when in fact it determined its estrangement from the land. Any land.

Architectural photography, independently but concurrently, consolidated this fractured view of architecture and landscape. It magnified its character as an independent object from the ground it rested on, instead of reinforcing its being an element inlaid into the macrostructure of nature. In pictures of buildings where landscape is also present, as they appeared in the 1930s in magazines, newspapers, and books, the outdoors is framed as a quiet visual commentary of the primary architectural statement. When modernism in architecture made photography a chief medium for its pictorial dissemination all over the world, the technological limits of black-and-white in representing greenery, and the default adoption of the standard sizes of the photographs of the era, had multiple side effects. First, it kept the viewer of the photograph out of the scene portrayed, excluding the individual from the experiential participation in nature. Second, it relegated vast areas of the sensorial context of landscape into the background of awareness. Third, it etched in the collective unconscious the proportional sizes of the picture frames as normative relational boundaries for seeing the environment. Fourth, architectural photography introduced into the mainstream media the idea of stylizing the landscape through the camera. In imposing a graphic order extraneous to the scene being captured by means of the viewfinder, it made the latent fluidity between architecture and landscape quietly elusive to the design professionals and the general public alike.

ABOVE:
Rockefeller Japanese Garden, Pocantico Hill, NY, 1978
Architect: Junzo Yoshimura
Landscape Architect: Unknown
Photographer: Ezra Stoller

OPPOSITE:
Cole House, La Habra, CA, 1959
Architect: Richard Neutra
Landscape Architect: Richard Neutra
Photographer: Julius Shulman

Pomerance House Swimming Pool, Cos Cobb, CT, 1938
Architect: Pomerance and Braines
Landscape Architect: Unknown
Photographer: Ezra Stoller

The early literature of this period cemented the differential treatment of buildings and sites as the new architecture took hold. Two examples representative of a prescriptive way of narrating architecture follow to illustrate this point. The first is Bruno Taut's *Modern Architecture*, published in 1929, one of the initial surveys gathering evidence of the robustness and practicality of the new idiom. It featured a gallery of completed achievements—a compelling mixture of engineering and architectural feats—but any reference to landscape architecture was conspicuously absent both in the pictures and in the text. The technology of their construction was the beginning of a fundamental change in architectural form interconnected to other novel silhouettes through urban and city planning, but unrelated to landscape. The space between the buildings was simply either the leftover of what remained between site boundaries and the building perimeter or a generic surface of undifferentiated character. The second is *The Modern House*, a book that the English practicing architect F. R. S. Yorke authored in 1934. It was a study of the critical changes the domestic environment had undergone since mechanization had taken over the materials, means, and methods of the building process, resulting in an entirely new architecture. Yorke thoroughly articulated the anatomy of the new residence in all its main parts. It placed the emergence of the Modern house in the context of a very short history of Modern architecture, recounted in the opening section of the publication. No mention whatsoever was made of the role of the garden in contributing to the development of a new way of dwelling in the world. In both books, widely popular when they were first published, the message was clear: the building is the focus; the garden is not.

Numerous practitioners engaged in writing along this vein about their own work or when wearing the hat of either the critic or the architectural historian, thus fortifying this fractured view of the environment. With this practice becoming dominant in broadcasting the spatial revolution, supplanting the traditional way of analyzing architecture through facade layouts and floor plan arrangements, the divorce between architecture and site became more or less permanent. It was a common custom then and retained a stronghold in subsequent years among architects and professional historians. Modernism put nature in the back seat of spatial discourse: it made nature irrelevant. The factory replaced the park, and the supreme attention devoted to the ideological and material implications of mass production deposed, figuratively speaking, the organic. The spatial revolution was underway, leaving landscape architects far behind. Their discipline found itself in the blind spot of architects traveling the modernist route at full speed. It disappeared from their rearview mirror.

ABOVE & OPPOSITE:
Charles O. Martin Residence, Aptos, CA, 1947–48
Architect: Clark & Beuttler
Landscape Architect: Thomas D. Church
Photographer: Rondal Partridge

Up until the rise of professionalism in the second half of the nineteenth century, the architect's scope of work encompassed structural engineering, albeit in empirical form, and landscape architecture, at least for its hardscape components. This gave diffused spatial cohesiveness to the built environment. Settlements in Western and non-Western traditions shared comparable attributes of spatial continuity between indoors and outdoors. Through a combination of the use of local materials and a grammar of proportionally compatible spaces, preindustrial urban complexes retained a remarkable consistency in design character. Separating out these subsets of design and relinquishing their realm of action to autonomous figures led to a more uneven environmental result. The price of lifting the fog of empiricism through the subdivision of design knowledge was a lack of thought consistency in the making of space. A loss of unity can be partially attributed to that event. Further, the introduction of new inventions in building technology in nineteenth-century society shaped

OPPOSITE:
Sequoyah House, Oakland, CA, 1955
Architect: Beverley D. Thorne
Landscape Architect: Robert Cornwall
Photographer: Philip Fein

ABOVE:
May House, Los Angeles, 1956
Architect: Cliff May
Landscape Architect: Thomas D. Church
Photographer: Julius Shulman

both process and outcomes, determining a new order of things. Concurrently, it de-skilled members of the building sector in the supplanted technology and dismantled the preceding organization of public and private space inside and out. This side effect of professionalism in design was a decisive contributor in the lack of understanding of the structural continuity between built form and landscape. Professionalism breeds specificity as opposed to integration of knowledge, marking deep lines of design separation at the juncture of an ultimately indivisible architecture and site.

In commenting on the split between architecture and the land, the noted landscape architect Garrett Eckbo, who rose to the professional scene in the late 1930s, commented:

> Gardens are outdoor spaces for people to live in. They therefore have to relate to houses, which are indoor spaces for living. To discuss either one alone is to discuss half of the picture, but because of the existing division of thought and labor we will have to do just that. Thought on gardens and houses has been divorced for so long that it is a little difficult to effect a reconciliation.[2]

ABOVE & OPPOSITE:
Gould Garden, Berkeley, CA, 1955–60
Architects: Gordon Reeve Gould with William Gillis and David Leaf
Landscape Architect: Lawrence Halprin
Photographer: Morley Baer

NORTH
LANDSCAPE PLAN
NANCY & REEVE GOULD
BERKELEY, CALIFORNIA
LANDSCAPE ARCHITECT

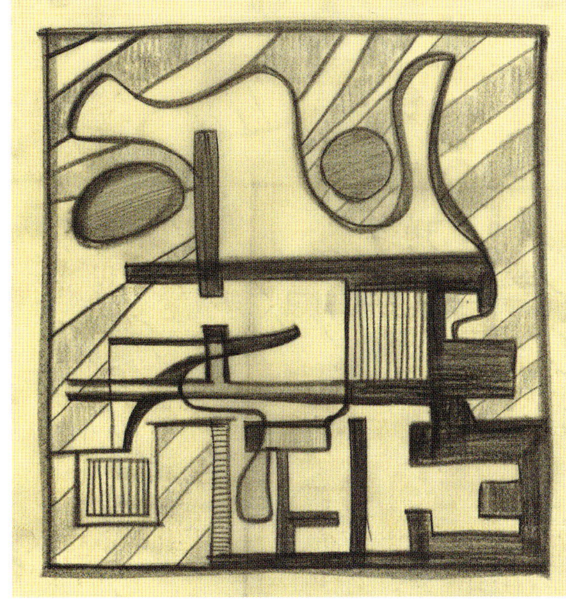

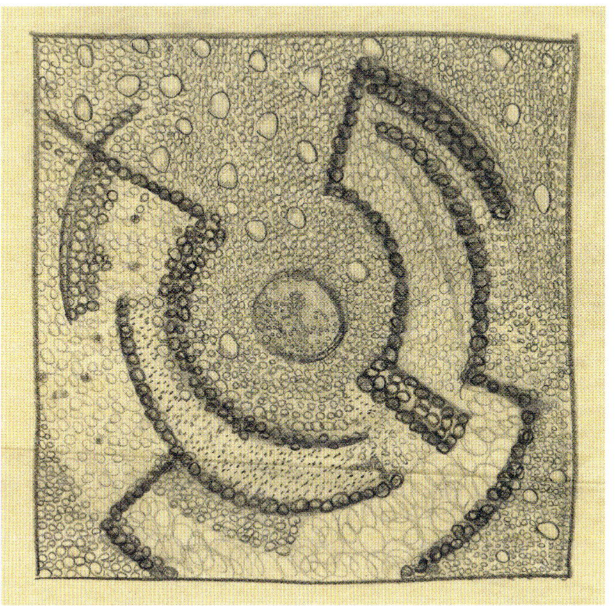

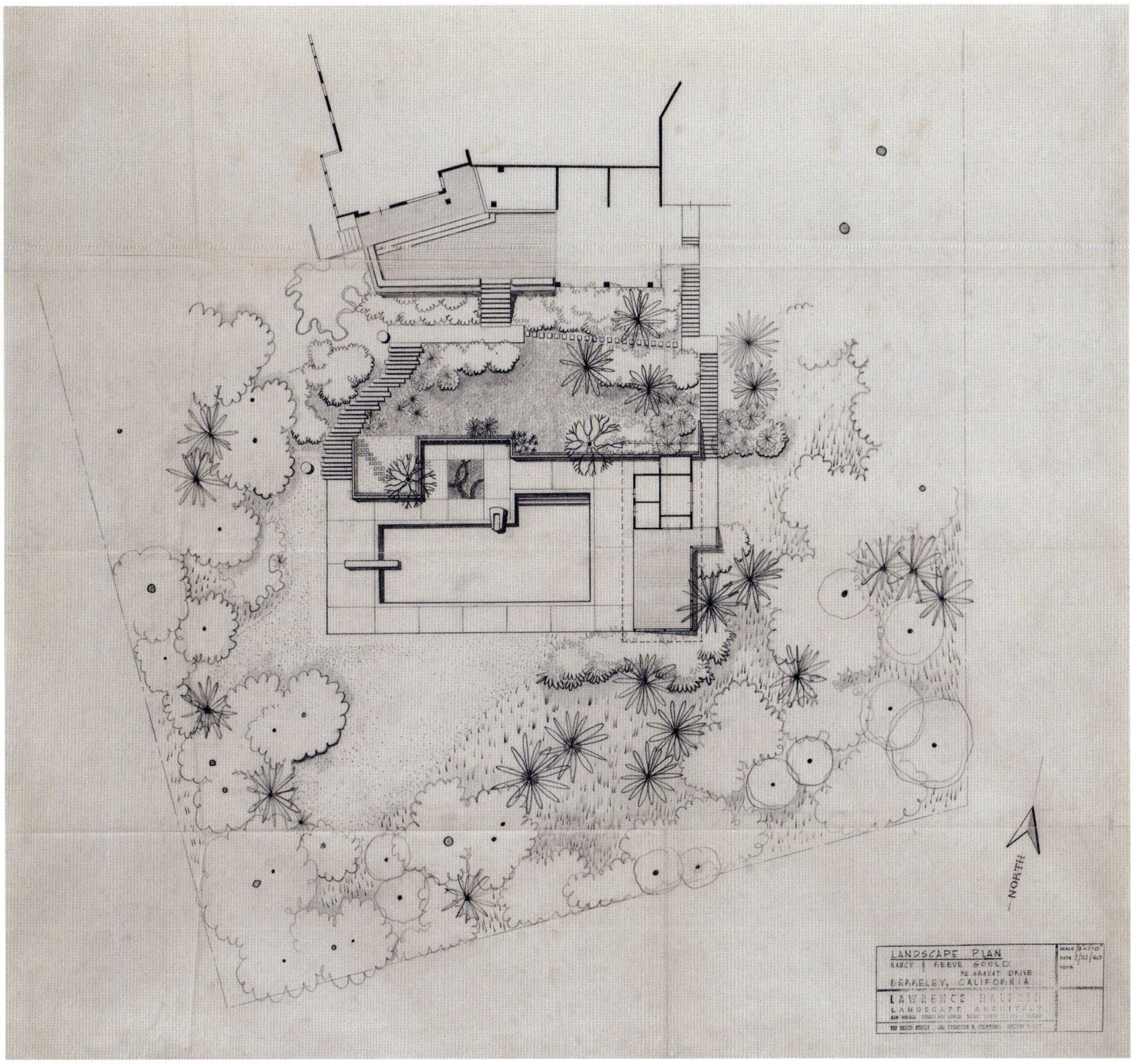

ABOVE:
Ladies' Home Journal Gardens, Location Unknown, 1949
Architect: Unknown
Landscape Architect: Unknown
Photographer: Ezra Stoller

OPPOSITE:
City Garden, Location Unknown, 1957
Architect: Unknown
Landscape Architect: Unknown
Photographer: Ezra Stoller

In fairness, the site underwent a fundamental redefinition as it related to the building in the worldview of the modernists: it became mechanized. By virtue of design being seen through the lens of mass production, the architecture of the land as a plastic statement of natural origins vanished to the eyes of the avant-gardists. The organic and the factory, figures of speech in a newfound polarity, were unlikely partners in the environmental revolution that was progressively being advocated on planet Earth. In its place, an infrastructural understanding of the terrain developed. The built environment underwent epoch-making changes in a matter of a few decades: electrification, forced water distribution, vehicular access through an ad hoc roadway system, security protocols due to recurring social unrests, fire safety priorities to attend to heightened risks in material combustion as building assemblies increased in complexity, and so forth. Dealing with the site turned out to be a technical affair, where competence and the pragmatics of use overruled all other concerns. The technological argument prompting the rhetoric of the modernists fell short in the debate on how to deal with landscape design, where the organic figures as a prominent agent in space formation.

ABOVE:
Connecticut General Life Insurance, Garage, Bloomfield, CT, 1972
Architect: Skidmore, Owings & Merrill
Landscape Architect: Unknown
Photographer: Ezra Stoller

OPPOSITE:
Holiday House, Quogue, NY, 1951
Architect: George Nelson
Landscape Architect: Unknown
Photographer: Ezra Stoller

OPPOSITE:
Geyer Terrace, Miami, FL, 1956
Architect: Unknown
Landscape Architect: Unknown
Photographer: Ezra Stoller

ABOVE:
Julian Saks Residence, Beverly Hills, CA, 1952
Architect: Paul Laszlo
Landscape Architect: Paul Laszlo
Photographer: Robert C. Cleveland

Another reason explains the landscape's invisibility. The conventional association between the garden and the romantic world of the previous century had evaporated, deemed irrelevant in a society shaped through the industrial ethos. The sentimental layer coated over the accounting of the landscape was a point of discomfort for the hardliners of the new architecture. Landscape was downgraded in architectural theory and practice because it was uninfluential in dealing with the social imperatives of attending to needs of an oppressed working class, and with the technological openings that the mass production of steel and concrete were offering to the practitioners committed to upgrading the built environment. In that moral dimension, any concentration on landscape architecture felt like a displacement of mental energy. Even Le Corbusier, in both his writings and his projects, gave a generic role to greenery in the development of the modernist city. When he provided access to open space at ground level to everybody by lifting buildings on *pilotis*, or stilt-like columns, he left the three-dimensional definition of those open spaces underneath and around the building simply *open*, that is without any formal direction on how they would relate to these new constructions beyond being free land for leisure activities of modern citizens.

Marko House, Coconut Grove, FL, 1956
Architect: Alfred Browning Parker
Landscape Architect: Unknown
Photographer: Ezra Stoller

Edward Kuzon House, Longmeadow, MA, 1961
Architect: Elroy Webber
Landscape Architect: Unknown
Photographer: Ezra Stoller

OPPOSITE:
Burnette Residence, Sarasota, FL, 1949–50
Architects: Paul Rudolph and Ralph Twitchell
Landscape Architect: Unknownl
Photographer: Ezra Stoller

ABOVE:
Julio Gallo Residence, Modesto, CA, 1951
Architect: Unknown
Landscape Architect: Douglas Baylis
Photographer: Rondal Patridge

Additionally, the soil as bearer of life occupied a dominant role in the ideological universe of the modernist architect, who saw the house as a healing environment from the ills the Industrial Revolution brought about across society. Biological references were increasingly entering the vocabulary of the avant-garde in an attempt to articulate a convincing link, to themselves and to others, between scientific discoveries and artistic achievements as contributors to architectural statements. Looking at foliage and vegetation was deemed to positively affect health, but that was the extent of that contrived observation, falling short of engaging in a more foundational debate on the interrelatedness of architecture and landscape. Both readings of the site were equally reductive, thus sealing its diminishing fate in architectural matters. The renowned historian Kenneth Frampton commented on the limits of these partial notions and on the actual potential of the site in being a determining factor in the development of space:

Rautbord House, Location Unknown, FL, 1965
Architect: Unknown
Landscape Architect: Unknown
Photographer: Ezra Stoller

> I used the term *organic* here to indicate a sensitively inflected approach toward the placement of the building in its site: one that synthesizes the sometimes conflicting demands of access, orientation, landfall, water table, prevailing wind, ecological imperatives and so on, without having immediate recourse to picturesque aesthetic effects.[3]

Structural engineering played its part as well in weakening the agency of the landscape on the final environmental outcome. It characterized the site as a diagrammatic plane on which to transmit loads coming from architecture. Members of that professional community scrutinized the soil composition for the sole purpose of establishing its load-bearing capacity for foundation design, magnifying the practical character of coming to terms with the existing conditions. The quest for a heroic and historically unprecedented verticality attained through the mathematical definition of structural sections was a core ambition fueling engineering thought, opposite to the prevailing horizontality of the landscape. The land was a medium to an end, an intermediary of negligible importance in the hierarchy of space formation.

In abstracting the massing of architecture, the early modernists abstracted the site as well. Even though every site is by default the material commitment to the specificity of place, the search for the universality of the architectural idea determined the erasure of the situated influence of the land and made ground zero of any particularity. Architects refrained from capitalizing on the site identity to feed enduring design statements and made architecture and city planning—both in their abstract figuration—the center of gravity of their philosophy. Architectural historian David Leatherbarrow remarked:

> Measure and extension are abstract characteristics of any site and while they can sustain the efficient and systematic execution of site plans, they can also prevent the designer from grasping any particular site's concrete qualities. In abstract space all sites are identical in kind as measures of extension and possess no uniqueness of position or orientation.[4]

Relinquishing any sustained attention to the design reciprocity between building and garden resulted in the loss of the preindustrial continuity between interior and exterior areas visible in any Western and non-Western settlements. The unifying thought that threaded together interior, building enclosure, hardscape, and landscape into a coherent whole gave way to the spatial fragmentation of the public realm and of the private sphere in the name

Broadwater Beach Hotel, Biloxi, MS, 1962
Architect: Gardner Stein Frank
Landscape Architect: Unknown
Photographer: Ezra Stoller

of progress. Historically, the fusion of hardscape and architecture through the architect's hand determined the seamless link between enclosed and open spaces. Gian Lorenzo Bernini's St. Peter's Square in Rome is an exemplary case of architecture reaching out into a landscape of stone, controlled dimensionally and spatially across scales. Even in Roman times, well before the establishment of the figure of the architect, this notion was present. Can architecture and landscape be separated in the settlement of Ghadamès in Libya, where the network of roads and dwellings are fused into one artifact? Instances of enmeshments of settings in nature abound in all those traditions. It was a diffused quality to be found virtually everywhere.

In his seminal book 1961 book, *Townscape*, British architectural writer and critic Gordon Cullen collected extensive proof of the granular level of design intervention that both ancestors and a few of his contemporaries brought into the architectural definition of urban voids. The vast majority of those examples looked at the ideas and themes feeding the development of macro and micro spaces that were generative of the characteristic connections found in historical cities. In Cullen's penetrating analysis of the grammar of open spaces, the massive scope of landscape architecture emerged in its full range. He detected and documented hundreds of architectural situations yielding spatial identity and design character in their concurrent existence. The choral sense of the city growing in sequenced and hybrid installments from different generations was a lasting message in the discipline.

Over time, however, a few landmarks of Modern architecture grew to be paradigmatic in redressing this uneven playing field and make it level once again. The Alden Dow Home and Studio in Midland, Michigan, is a textbook example of an environmental totality where the garden and the house are inseparable. It is unimaginable today to experience the park towers, town houses, courthouses, and all the accessory buildings of Lafayette Park in Detroit, by Mies van der Rohe and Ludwig Hilberseimer, without Alfred Caldwell's commanding landscape design. The interpenetration of architecture and nature is the core idea behind the Oakland Museum of California, a milestone in the career of Roche & Dinkeloo. The assertive choreography of the open spaces by landscape architect Dan Kiley determined the forms of an inhabited artificial topography. In Italy, Carlo Scarpa's design for the Brion Vega Cemetery a few miles out of Treviso arranges its archaic forms in a precinct where the gentleness of the meadow provides the counterpoint to the roughness of the cast-in-place concrete.

Positioning and orienting architecture is a primary act, which is inherently a landscape gesture. The land always precedes the architecture. It comes with its own spatiality before that imposed on its soil through human intervention. The site exerts pressure

Ketti Frings House, Los Angeles, CA, 1959
Architect: Unknown
Landscape Architect: Unknown
Photographer: Ezra Stoller

Georgi House, Pasadena, CA, 1954
Architect: Boyd Georgi
Landscape Architect: Eric Armstrong
Photographer: Julius Shulman

THIS PAGE & OPPOSITE:
**Heymes Residence,
San Francisco, CA, 1948**
Architect: John Funk
Landscape Architect: Douglas Baylis
Photographer: Roger Sturtevant

on the architects on the sheer basis of its material properties and its spatial identity. Yet architectural manuals present site-less architecture, a true contradiction in terms since architecture is always situated. Can anyone make a credible case that Louis Kahn's Salk Institute for Biological Studies in La Jolla, California, could be located on a different site from where it is? Would the open court make any sense in the absence of that horizon toward the Pacific Ocean, as well as the distance between buildings so that anyone can appreciate the bold outline of its plan from all sides? Whether a masterpiece or an average building, they both come to terms with the site in a major way.

The disjointed nature of planning house and garden sequentially had a turning point in the 1930s. Toward the end of the decade, the relationship between the two was revisited, and the result significantly expanded the very notion of what a garden was from previous definitions. Garden as space was a novel concept in that decade, yet slowly coming into being. Eckbo wrote:

ABOVE:
Otto Spaeth House, Southampton, NY, 1957
Architects: Gordon Chadwick and George Nelson
Landscape Architect: Karl Linn
Photographer: Ezra Stoller

OPPOSITE:
Keefer Residence, Location and Date Unknown
Architect: Unknown
Landscape Architect: Robert Royston
Photographer: Rondal Partridge

ABOVE AND OPPOSITE:
Upjohn, Kalamazoo, MI, 1961
Architect: Skidmore, Owings & Merrill
Landscape Architect: Sasaki, Walker and Associates (SWA)
Photographer: Ezra Stoller

> Although people live on the land, they live in the space immediately above it; this is simply a volume of air of appropriate size and form. We are becoming more and more sure that the form given this block of space is of controlling importance. This involves a sculptural conception, rather than an arrangement of objects in a vacuum.[5]

Eckbo intuited the richness in the cross-pollination between the two. On one side, the garden could be "architecturalized," such as in Robert Royston's Chinn Garden in San Francisco, completed in 1950, and architecture could take on aspects of the natural realm. Nature has come inside the house, as in the 1945 Colet Residence in Lincoln, Massachusetts, by Carl Koch, and in the 1954 Knauer Residence, Los Angeles, designed by Rodney Walker; and the house has gone out in the landscape, such as in the 1962 McIntyre Residence, Hillsborough, California, designed by Lawrence Halprin.

Eckbo continued: "The formal garden forces architecture upon the landscape; the informal garden forces that garden upon the architecture."[6]

The resonance of materials and forms from the garden into the house and from the house into the garden was a critical extension in landscape architecture of the indoor-outdoor theme ubiquitous in residential architecture from the 1920s on.

The plastic nature of garden space and the idea that the garden starts from the house started taking hold. Seeing the ground as something more than an ecological resource, or a naturalistic asset to revel into, was a mandatory step in having landscape architects enter the design areas as the latest protagonists in space making. Human response to smaller spaces started being paid attention to. But the forms of the garden are never quite fixed, as growth is an inherent force in the natural process of perpetual change. How to combine the irregular forms of nature with the assertive forms of Modern architecture was the next challenge in arriving at an integrated whole.

Joseph Hudnut, first dean of the Harvard Graduate School of Design and instrumental in bringing Bauhaus founder Walter Gropius to the school in 1937, wrote about this new conception of space:

> The new vision has dissolved the ancient boundary between architecture and landscape architecture. The garden flows into and over the house: through loggias and courts and wide areas of clear glass, and over the roofs and sun-rooms and canopied terraces. The house reaches out into the garden with walls and terraced enclosures that continue its rhythms and share its grace. The concordant factor is the new quality given to space.[7]

OPPOSITE:
**Shapiro Screened Porch,
Location Unknown, FL, 1955**
Architect: Kenneth Jacobsen
Landscape Architect: Unknown
Photographer: Ezra Stoller

THIS PAGE:
**Pellissier Residence,
Whittier, CA, 1955**
Architect: Unknown
Landscape Architect: Garrett Eckbo
Photographer: Maynard Parker

Castle House, New London, CT, 1961
Architect: Ulrich Franzen
Landscape Architect: Ulrich Franzen
Photographer: Ezra Stoller

When these words were being written, the nineteenth-century notions of the garden were still lingering well in the following century. The pleasure garden—garden as escape, as refuge, as private paradise, or as a locus for major recreational purposes—were reminders of ephemeral charm, lacking in architectural substance. The urge to connect landscape design to modernity grew rapidly in consonance with the all-encompassing changes that movement brought to society.

Eckbo was among the first who asked key questions with a new perspective:

> First, what do we mean by "landscape architecture"? In the broadest definition the landscape is everything one can see from a given station point or series of them, and thus is compounded by building and landscape development.[8]

Bruce Alspach House,
Location Unknown, FL, 1962
Architect: Alfred Browning Parker
Landscape Architect: Unknown
Photographer: Ezra Stoller

In this more inclusive definition, he was raising the figure of the landscape architect to the level of the master builder. It was a jump in scale, in mandate and agency, a position negated until then regarding the landscape architect. In a later article, Eckbo spoke of architecture as "shelter development"[9] overturning the dominance of architecture over landscape, by making architecture a subset of landscape. Landscape as an appendix to architecture vis-à-vis the notion of architecture as an appendix to landscape was being entertained under these new conditions. In practical terms, the conception of the garden radically changed in those years and opened new vistas to modernist architects as well. Gardens could exist on human-defined planes altogether different from those available in natural settings. Roof gardens, such as the Kaiser Roof Garden in Oakland, California, designed in 1960 by Theodore Osmundson for the Kaiser Center, by Welton Becket, became powerful event spaces. But even in smaller

Kaiser Roof Garden, Oakland, CA, 1960
Architect: Welton Becket
Landscape Architect: Theodore Osmundson
Photographer: Theodore Osmundson

IBM T. J. Watson Research Center, Yorktown, NY, 1961
Architect: Eero Saarinen
Landscape Architects: Hideo Sasaki and Stu Dawson
Photographer: Ezra Stoller

OPPOSITE:
Fitch House, Stony Point, NY, 1950
Architect: James Marston Fitch
Landscape Architect: Unknown
Photographer: Ezra Stoller

ABOVE:
Simon Roof Garden, San Francisco, CA, 1950–51
Architect: Unknown
Landscape Architect: Lawrence Halprin
Photographer: Ernest Braun

versions such as in the Simon Roof Garden in San Francisco, designed and completed between 1950 and 1951 by Lawrence Halprin, they vastly increase their influence in the overall experience of architecture.

Gardens started entering the buildings and determining spaces within spaces. The Jaret House, Hollywood, of 1959, by Buff, Straub & Hensman; Fitch House, Stony Point, New York, of 1950, designed for architectural historian James Marston Fitch; and the Marin County Civic Center in San Rafael, California, of 1962, by Frank Lloyd Wright, are three illustrative examples of such garden conditions. Balconies; verandas; decks, such as the 1959 Montez House in Oakland, California, designed by Roger Lee; open porches; and outdoor rooms, like that located at the roof level of the 1947 Kaufmann Desert House in Palm Springs, California, by Richard Neutra, all became extensions of nature into architecture.

Landscape architecture calls for the crafting of spatial events. Whether stand-alone sites, intertwined with the architecture, or treated as a quiet commentary to the latter, landscapes always exist and precede the arrival of architecture. Within the landscape continuum, gardens are bounded portions of the landscape, their roof always being the sky and the ground always being the canvas of the landscape architect. Within the limits of the garden and those of the house, dynamic relationships were newly established, triggering imaginative production by design creatives as they were perceiving the spatial potential of distinct environmental conditions. Visual and material resonances between the architectural lines of the building and the natural lines of the garden started appearing in that period. They offered great opportunities to the postwar generations of landscape architects to experiment

ABOVE:
Jared House, Los Angeles, CA, 1959
Architect: Buff, Straub & Hensman
Landscape Architect: Emmet Wemple
Photographer: Ernest Braun

OPPOSITE:
Montez House, Oakland, CA, 1959
Architect: Roger Lee
Landscape Architect: Unknown
Photographer: Ernest Braun

Kaufmann Desert House,
Palm Springs, CA, 1947
Architect: Richard Neutra
Landscape Architect: Richard Neutra
Photographer: Julius Shulman

ABOVE:
Weisman House, Los Angeles, CA, 1963
Architect: Rick Faber
Landscape Architect: Unknown
Photographer: Julius Shulman

OPPOSITE:
Colet House, Location Unknown, 1945
Architect: Carl Koch
Landscape Architect: Unknown
Photographer: Ezra Stoller

and cut their teeth in preparation for larger works. Through a late awakening of Modernism, from a two-dimensional exercise in marking the ground, landscape architecture became a three-dimensional affair. To this point, landscape architect Douglas Baylis delivered a dual approach through geometric landscape and hardscape in the garden for the Heymes Residence in San Francisco, completed in 1948, in an architectural design by John Funk. The Edward Kuzon House in Longmeadow, Massachusetts, completed in 1961, is another example. Many others followed suit.

That architecture and landscape were one had a consequential spokesperson in Frank Lloyd Wright, who wrote in one of his last books before his passing:

ABOVE AND OPPOSITE:
Green Johnson House, Mill Valley, CA, 1962
Architect: Marquis & Stoller
Landscape Architect: Royston, Hanamoto, Mayes and Beck
Photographer: Ezra Stoller

Architectural features of any democratic ground plan for human freedom rise naturally by, and from, topography. This means that buildings would all take on, in endless variety, the nature and character of the ground on which they would stand and, thus inspired, become component parts. Wherever possible all buildings would be integral parts—organic features of the ground—according to place and purpose.[10]

Footnotes

1. Joseph Hudnut, "The Modern Garden," in Christopher Tunnard, *Gardens in the Modern Landscape* (1938; rev. ed., London: Architectural Press; New York: Charles Scribner's Sons, 1948), 178.
2. Garrett Eckbo, "Outdoors and In: Gardens as Living Space," *Magazine of Art* 34, no. 8 (October 1941), 422.
3. Kenneth Frampton, "In Search of the Modern Landscape," in *Denatured Visions: Landscape and Culture in the Twentieth Century*, ed. Stuart Wrede and William Howard Adams (New York: Museum of Modern Art, 1991), 43–44.
4. David Leatherbarrow, *The Roots of Architectural Invention: Site, Enclosure, Materials* (Cambridge: Cambridge University Press, 1993), 17.
5. Eckbo, "Outdoors and In," 422.
6. Ibid., 425.
7. Hudnut, "The Modern Garden," 178.
8. Garrett Eckbo, "What Do We Mean by Modern Landscape Architecture?," *Journal, Royal Architectural Institute of Canada* 27, no. 8 (August 1950): 268–69.
9. Garrett Eckbo, "Architecture and the Landscape," *Arts & Architecture*, October 1964, 22.
10. Frank Lloyd Wright, *The Living City* (New York: Horizon Press, 1958), 112.

Edells Garden, West Portal District, San Francisco, CA, 1954
Architect: Unknown
Landscape Architect: Robert Cornwall
Photographer: Rondal Partridge

DAY-LITE MARKET

PART 3

Landscape Architect and Architect: A Complex Relationship

Architecture and Acreage Together are Landscape.[1]

FRANK LLOYD WRIGHT

SITE IS THE DEFAULT preexisting condition of architecture. It always precedes it. The site is that environmental asset shared equally between landscape architects and architects. This common bond starting from two different points of departure, the land and the building, has a meeting point in the middle of this uniting thread: the total environment. Although this is a given, its self-evidence strangely slipped into the background as Modern architecture came into being.

The reciprocal awakening of landscape architects and architects about their own interrelatedness and unity was gradual. Over time, they came to terms with each other through joint incremental realizations, each camp mining the other's sources to establish a common language under the auspices of an inevitable modernity. For both communities, space was the mutual nucleus of their parallel endeavors. The traditional conception of gardens as images to be admired at a distance gave way to gardens as settings promoting immersive experiences with minimal navigational script. Their traditional fixity was deactivated through the adoption of a novel conception of space discovered dynamically through meandering as a first design principle. For their part, buildings expanded their agencies beyond the material precincts of their floor plans and spilled into the land. Walls started sliding out of their plinths, suggesting nested spaces of different scales, some covered, some out of doors.

Broadly speaking, together with site, space is therefore another common bond that irreversibly joints landscape architects and architects. Conceptually and materially, they both concurrently lay claim on space. Its fluidity makes any attempt from either faction to enclose it and control it unilaterally elusive, an artifice of sorts. Any opening, whether a door, a window, or a skylight, is an instance of the porosity of space, intractable to trapping. The notable architect Louis Kahn aptly wrote, "A room is not a room without natural light,"[2] establishing a direct connection between the coming into being of architecture through the making of the room and the permanent link to the surrounding landscape. The architectural act is in and part of the larger environment that contains it, no matter what. The piercing of the building enclosure is the spilling over of the cotained void into its adjacent open void, sanctioning its indivisibility. Space can be neither subdivided nor sealed. As Modern architecture increased the size of the openings, making itself permeable to outside influence, that spatial continuity became patently obvious: it gave itself to the landscape just as much. The attributes of this new conception of space were to be explored through matter-of-fact design interventions in indoor and outdoor conditions in the ensuing decades.

The Industrial Revolution brought faster movement as a factor to consider in the organization of space. As mass society was put in motion, the viewpoint of each of its members picked up speed.

OPPOSITE & PREVIOUS SPREAD:
Ernest Meyer and Charles B. Kuhn Residence, Aptos, CA, 1948
Architect: Unknown
Landscape Architect: Thomas D. Church
Photographers: Maynard Parker (p. 119) and Rondal Partridge (p. 120)

Wright Ludington House,
Santa Barbara, CA, 1960
Architect: Bertram Goodhue
Landscape Architect: Unknown
Photographer: Ezra Stoller

The perception of space through accelerated navigation yielded a feeling in the citizens of this emerging society that had no historical precedent in previous millennia. No longer was the pace of the walk or the carriage the medium for the kinetic experience of space. The rapidly changing perspective through which the new landscape was being consumed made the fixity of the vanishing point of Renaissance memory anachronistic and irrelevant to landscape architects as the twentieth century was marching on. Architects understood and capitalized on this realization to make their case for a novel environment reflective of these new patterns of occupation, while remaining silent on the question of shaping a suitable landscape for modernity. Their utmost attention was on reforming the building from the ground up, rather than the site that that same building was part of, as a matter of course.

Mies van der Rohe rarely commented on this topic since, by his own admission, his singular preoccupation was building, both the thought behind it and its ensuing outcome. Yet in one of his most circulated interviews, which was recorded in 1955 at the Waldorf Astoria hotel in New York, he made two critical remarks. The first was tied to the influence of Frank Loyd Wright on his work: "I think more as a liberation, you know, I felt much freer by seeing what he did; the way how he put the building in the landscape, and the free way, how he uses space and so on."[3] The second was related to the color treatment of the steel frame of his most notable residential landmark in Plano, Illinois: "At the Farnsworth's house I painted it white, because it was in the green, it was in the open. I could use any color, you know."[4] Regarding the first reference, the German master acknowledged that the "how" to place his structure in landscape was in fact a foremost architectural problem. As for the second, Mies made a design decision based on the surrounding greenery, to paint the steel frame a different color from its customary black. For Mies, structure was the basis for space making. The landscape, for him, had no structure in an architectural sense. He respected engineers as skilled accomplices in actualizing his resolve to develop space out of a structural idea, while he left no enduring statement about the role of landscape architects.

When Mies painted the steel of the Farnsworth House white in response to the landscape, he was not intervening actively in the landscape, only indirectly. Still, in being reactive to it, he recognized its agency on the making of a building, inseparable in their simultaneous presence. In the same 1955 interview, an example of landscape came up as he described a reading of details versus overall form while discussing his design in the creation of the Seagram Building on Park Avenue in New York:

OPPOSITE:
Arkin Garden, San Francisco, CA, 1955
Architect: Unknown
Landscape Architect: Robert Cornwall
Photographer: Rondal Partridge

ABOVE:
Goldstone House, Beverly Hills, CA, 1948
Architect: Unknown
Landscape Architect: Eckbo, Royston & Williams
Photographer: Julius Shulman

> It is not an individual thing, you know—thousands of windows, good or bad, that doesn't mean anything, it is like an army of soldiers, or like a meadow, you don't see the details anymore when you see the mass, I think that is the quality of this tower.[5]

He was bringing up the issue of scale in landscape. In zooming out of the meadow, the eye stops seeing its details while apprehending the mass. Leveraging this insight, he organized the space on the ground between his buildings by layering floating platforms; kept them level as much as possible; and made them of extremely durable materials so as to have its image fixed in philosophical perpetuity, despite use and abuse. His are examples of gardens made entirely of hardscape.

Bauhaus founder Walter Gropius remarked on this issue from the professional angle of the division of labor in design, in another widely circulated interview:

> I am very much against these artificial boundaries between landscape architecture and architecture and exterior architecture and interior architecture and so on, because the principles are all the same for everything of our surroundings, you know. So, it's left to the individual what he is most interested in, but I have been dragged into many directions myself, you know, I tried this and that because I was interested in it, you know.[6]

Although these statements were made in the postwar period, it is legitimate to wonder why, for example, landscape architects were notably missing from the first conference of the Congrès Internationaux d'Architecture Moderne (CIAM), the International Congresses of Modern Architecture, which was organized at the Château de La Sarraz, Switzerland, in 1928, nearly a decade after the founding of the Bauhaus. Of all the architects who participated in that venue, the only one with a previous

ABOVE:
Convair Astronautics, San Diego, CA, 1958
Architect: Pereira & Luckman
Landscape Architect: Unknown
Photographer: Julius Shulman

OPPOSITE:
Capital Park Apartments, Washington, DC, 1959
Architect: James H. Scheuer
Landscape Architect: Unknown
Photographer: Ezra Stoller

Central Motor Bank,
Jefferson City, MO, 1962
Architect: Skidmore,
Owings & Merrill
Landscape Architect: Unknown
Photographer: Ezra Stoller

direct connection to landscape design was Gabriel Guevrekian, the conference organizer together with Le Corbusier. This singular figure was born in Turkey of Armenian parents, raised in Tehran, and educated in Vienna, before moving to Paris after graduation. He designed groundbreaking Cubist gardens, first in the 1925 Exposition Internationale des Arts Décoratifs et Industriels Modernes, in Paris, and then, shortly after, inlaid at the Villa Noailles, Hyères, in southern France, which Robert Mallet-Stevens designed in the 1920s. These achievements were widely circulated at the time in the compressed black-and-white photographs published in U.S. magazines that American landscape architects were reading, paving the way for an extension of the modernist treatment

Francis Kirkham Residence,
San Francisco, CA, 1948
Architect: Unknown
Landscape Architect: Thomas D. Church
Photographer: Rondal Partridge

**Knauer Residence,
Los Angeles, CA, 1954**
Architect: Rodney Walker
Landscape Architect: Unknown
Photographer: Julius Shulman

of space into the garden. However, Guevrekian's training was as an architect, not a landscape architect; he took on landscape projects as a subset of his design activities. Among those participating architects at the first CIAM conference, others also undertook landscape design projects around the same time or later. For their own residences, Ernst May in Frankfurt in 1928 and André Lurçat in Paris in 1930 tackled the design of a garden using a common formal language: exporting the geometry of the architecture for making the outdoor spaces.

Le Corbusier did bring the garden into the emerging modernist discourse in his famous *Les 5 points d'une architecture nouvelle*, published in 1926. The second point was *les toits-jardins*,

THIS PAGE & OPPOSITE:
Connecticut General Life Insurance, Bloomfield, CT, 1957
Architect: Skidmore, Owings & Merrill
Landscape Architect: Isamu Noguchi
Photographer: Ezra Stoller

literally "the roof gardens," thus assigning some imprecise role to the organic in the development of the new image of architecture. Throughout his projects, placeholders for greenery were consistently drawn in his wire-frame renderings, plans, and elevations. These opportunities for vegetation, however, were offered to the reader in his books and articles with neither a specific attitude toward their spatial treatment nor a credible indication of his having any knowledge of plant materials. While Villa Church in Ville-d'Avray of 1927–29 and the Beistegui Rooftop in Paris of 1930 are among two of the best-known examples of active landscape design by Le Corbusier and his cousin Pierre Jeanneret, landscape architecture had a decisively lesser role in his philosophy and output compared to that of Frank Lloyd Wright, who made nature the generative center of all space making. For Wright, architecture was a spatial engagement with the site in the spirit of an indissoluble unity.

Besides site and space, a third common denominator between architecture and landscape specific to the first part of the twentieth century was the influence of painting on their

mutual future course. The triangulation of painting, landscape, and architecture was a modernist constant in the development of a new vision of space. Painting opened up a visual universe unfettered from the practicality of architecture, and it largely anticipated spatial possibilities later explored three-dimensionally. A painting is always bounded just as a garden and a piece of architecture are. As opposed to open landscapes, gardens have beginnings and ends. Within their confined areas, gardens will expand until they encounter a barrier, whether material and legal, that will outline its resulting boundaries when all elements are combined. Together with the epochal technological changes that occurred throughout the nineteenth century, the artistic revolution in painting jolted the formal certainties of architecture at its foundation, providing a rich source of formal flights at the center of a watershed of unprecedented imagery. Architecture first, and landscape architecture subsequently, heavily capitalized on these momentous shifts to determine unmatched environmental conditions in their respective domains for a society undergoing constant reinvention. The cross-pollination involving painting,

sculpture, and architecture reached its apex between the 1920s and the 1930s, with Paris as its symbolic capital. Landscape architecture largely benefited from that interdisciplinary exchange in the making, since many forms later to be found in gardens grew out of the formal world of painting. The kinetic experience of the garden came into play. With the Modern garden, the possibility of experiencing a painting three-dimensionally opened up in a way that was inaccessible to architecture framed within functional and structural limits. It offered concurrently the fusion of function and of figurative and plastic space.

The counterintuitive perceptual hierarchy of building and landscape, the former dominant in photographic representation and the latter the protagonist in the real-life experience of space, constituted a conundrum with which landscape architects trained in the Beaux-Arts tradition were coming to grips. Architects became progressively aware of the environment beyond their buildings, and landscape architects became increasingly conscious of architecture's role in the joint construction of the environmental experience.

ABOVE:
Stockstrum House, St. Louis, MO, 1953
Architect: Harris Armstrong
Landscape Architect: Thomas D. Church
Photographer: Ezra Stoller

OPPOSITE:
Smalley House, Los Angeles, CA, 1974
Architect: A. Quincy Jones
Landscape Architect: Unknown
Photographer: Julius Shulman

THIS PAGE & OPPOSITE:
Caribe Hilton, San Juan, Puerto Rico, 1951
Architect: Toro-Ferrer y Torregross
Landscape Architect: Unknown
Photographer: Ezra Stoller

Each group started writing in favor of a fusion between the two, like teams drilling the rock of a tunnel from opposite ends. In airing their aspirations, as well as frustration and discontent, landscape architects started writing in architectural magazines, *Pencil Points*, the *Architectural Forum*, and the *Architectural Review* among them.

James Rose, a young Harvard student expelled from the program for his unwillingness to follow the untenable Beaux-Arts precepts in the late 1930s, wrote a seminal article in which, among other things, he raised a key question encapsulating the crux of the landscape architecture/architecture problem:

> Isn't it a little inconsistent, and perhaps unfair, to expect a Twentieth-Century individual to step out of a stream-lined automobile and then flounder through a Rousseauian wilderness until he reaches a "machine for living"?[7]

The story of the three student rebels—Dan Kiley, Garrett Eckbo, and James Rose—out of the body of students attending the landscape

architecture program at Harvard from 1936 on is an integral part of the history of the discipline in its modern phase. That the three of them crossed the institutional lines and started taking courses at the Harvard Graduate School of Design at the time of Walter Gropius's arrival in the autumn of 1937 signaled that the next generation of landscape architects had begun asking architects for a way out of their impasse in order to explore three-dimensional space free of any historicist shadow. Young students of landscape architecture went to architects to take that quantum leap into modernity, seeking the forging of a unified vision of the house and garden, a symbolic theme at the cellular level for the reformation of the built environment. The principles of its dynamic occupation of space that the Bauhaus founder and his comrades advanced in the school catapulted landscape architecture into an experimental trajectory opening one of the most creative chapters in the field. In doing away with the locked geometries of the Beaux-Arts aesthetic, their updated focus was the design exploration of a fluid space in perpetual perceptual motion, yielding concurrent multiple perspectives of equal three-dimensional relevance. Landscape architects were eager to shed any associations of their design territory with the period of the picturesque, so as to embrace the principles of space making in modern times. Organic and

ABOVE:
Irving Levin Garden, San Anselmo, CA, 1950
Architect: Unknown
Landscape Architect: Lawrence Halprin
Photographer: Ernest Braun

OPPOSITE:
Lyons House, Berkeley, CA, 1957–58
Architect: Joseph Esherick
Landscape Architect: Thomas D. Church
Photographer: Ezra Stoller

THIS PAGE & OPPOSITE:
Marco Wolff House, Los Angeles, CA, 1959
Architect: Ladd & Kelsey
Landscape Architect: Thornton Ladd
Photographer: Ezra Stoller

OPPOSITE:
Rangell House, Los Angeles, CA, 1952
Architect: Unknown
Landscape Architect: Garrett Eckbo
Photographer: Julius Shulman

ABOVE:
Dillon House, Location Unknown, 1953
Architect: Unknown
Landscape Architect: Unknown
Photographer: Julius Shulman

inorganic materials provided the full palette to mold an outdoor void replete with the range of three-dimensional experiences that architects were forwarding in the shaping of the built environment.

It will take some time before concrete concerns about the ecology and the increasingly compromised balance of the ecosystem, possibly the most important contribution of Ian McHarg, rise above the ideological horizon of a nascent environmental conscience, leading to a socially responsible definition of the goals of landscape architecture. Lists of material and geological assets, as McHarg was advocating, became dominant tools of analysis at the macro scale of the territory and the region, with consequent losses of the understanding of space at the scale of the individual. At this stage, the conversation between landscape architecture and architecture was primarily a strictly artistic, formal affair.

This overture toward architecture in the landscape circles continued for several more years. A decade later, the notable landscape architect Lawrence Halprin studied under Gropius and

Marcel Breuer as well as with Christopher Tunnard, who had just arrived from England following his groundbreaking texts on landscape architecture and modernity. Figuratively speaking, architecture fed landscape architecture, then in an ailing state. The mechanization of society had obliterated the relevance of nature in the philosophical fundamentals of human experience shaped through modernity. The question then remains: why did landscape architects seek architects to emerge from the quicksand of historicism, when those same architects had nothing to say about nature?

It is in this intense decade that, through the copious writings of primarily Tunnard, Eckbo, and Rose, the course of landscape architecture in the United States was reset on modernist tracks. Once landscape architects turned the tide and embraced the Modern revolution, they asked architectural photographers, the same image makers creating the visual inevitability of the

ABOVE:
Hixsons House, Pasadena, CA, 1952
Architect: John Matthias, designer
Landscape Architect: Unknown
Photographer: Julius Shulman

OPPOSITE:
Macht House, Baltimore, MD, 1956
Architect: James Rose
Landscape Architect: James Rose
Photographer: Ezra Stoller

Gayer House, Coconut Grove, FL, 1956
Architect: Alfred Browning Parker
Landscape Architect: Unknown
Photographer: Ezra Stoller

new idiom in architecture, to point their cameras toward outdoor architecture and include them in the master narrative of modernity in design. In this revolt against historicism, landscape architects endeavored to translate the same concepts that their architect colleagues theorized into a landscape version. Modular planning had modular gardens, the open plan had the open garden, the loss of symmetry had the circuitous landscape, and so forth.

Together with painting, sculpture gathered momentum in becoming a reference to reshape the garden in modernist terms. For some, landscape architecture could be designed as an immersive sculpture. James Rose wrote:

> You can walk through it. You are inside something. You have to feel that you are inside something, even though you are

Morris House, Armonk, NY, 1949
Architect: Huson Jackson
Landscape Architect: Unknown
Photographer: Ezra Stoller

out of doors, instead of being outside of something trying to think everything else away. A garden is a sculpture from any place you are in it, even while you are in motion and there's nothing outside that has to be thought away because that's part of it too—just as you are.[8]

Therefore, it is something to be part of rather than something to look at. Of the three mavericks, Rose appeared the most determined to dispose of the predetermined geometry, existing independently from human occupation and inhabitation, that historicism had caged landscape architecture in. He advocated for an open structure pregnant of countless possibilities for spatial discovery. Rose's emphasis on garden as sculpture was a theme to which he stayed true for the rest of his life.

OPPOSITE:
De Sabla Apartments, San Mateo, CA, 1951
Architect: Angus McSweeney
Landscape Architect: Angus McSweeney
Photographer: Rondal Partridge

ABOVE:
Schulman House, Akron, OH, 1961
Architect: Unknown
Landscape Architect: Unknown
Photographer: Ezra Stoller

Other developments were taking place on the West Coast. The friendship between the veteran landscape architect Thomas Dolliver Church and architect William Wilson Wurster, in the same late thirties, was particularly revealing of a working relationship that seamlessly went from the personal to the professional. They traveled together, already seasoned practitioners, to the Nordic countries to visit sites and meet contemporaries working in the new architecture. Without any notice, they knocked on the front door of Alvar Aalto's newly finished personal residence in the Munkkiniemi district of Helsinki, at the time when he and his wife, Aino, had on the drawing boards the Villa Mairea in Noormarkku, Finland, a project featuring, among many other innovations, the renowned

pool, later popularized as "kidney pool" in North America, while for the Aaltos that shape, realized in 1939, was a metaphor of the Finnish Lake. Wurster wrote a decade later: "Architecture and landscape architecture are one thing when they are separated only as to materials and techniques, not as a basic approach."[9]

Through Aalto, Church—a Harvard-trained landscape architect in the Beaux-Arts pedagogy, designing Italian and Spanish gardens until then and who would be the representative for Artek furniture on the West Coast for a number of years—directly experienced the outreach of architects into the landscape and how that integration could actually occur. The organic form entered assertively into his design vocabulary. This unplanned meeting had profound implications on the careers of both creatives and on the dissemination of

ABOVE:
Emmons House, Pacific Palisades, CA, 1957
Architect: Jones & Emmons
Landscape Architect: Eckbo, Royston & Williams
Photographer: Julius Shulman

OPPOSITE:
Koch House II, Concord, MA, 1953
Architect: Carl Koch
Landscape Architect: Unknown
Photographer: Ezra Stoller

OPPOSITE:
Bauer House, Hope Ranch, CA, 1958
Architect: Carl Maston
Landscape Architect: Unknown
Photographer: Julius Shulman

ABOVE:
Engelberg House, Los Angeles, CA, 1949
Architect: Henry Robert Harrison
Landscape Architect: Eckbo, Royston & Williams
Photographer: Julius Shulman

Hunting Lodge, Sterling City, TX, 1966
Architect: Frank Welch
Landscape Architect: Dan Heyn
Photographer: Ezra Stoller

this idea—the fusion of architecture and landscape into a coherent environmental whole—around the United States, most certainly in the Golden State. The Donnell Garden in Sonoma, California, completed in 1947, is the pinnacle of that possibility and a transformative milestone in the genre that echoed around the world. A beginner, Lawrence Halprin, was hired shortly after graduation and before the design of the garden started, thus exposing the next generation to that possibility as well. As part of his staff, Church also employed two young architects, George Rockrise and Germano Milono, reversing the traditional dominance of architecture over landscape in the composition of the design team.

In this period, Church started developing gardens with an increasingly architectural character, such as in the Charles O. Martin Residence and Ernest Meyer and Charles B. Kuhn Residence, both located in Aptos, California, and completed around the same years. The primary photographer of this phase in Church's career was one of the sons of the world-renowned photographer Imogen Cunningham, Rondal Partridge, who, following his apprenticeship with Edward Weston and Dorothea Lange, would bring forth a fine arts photographic rendition of a new landscape architecture. A historicist turned modernist landscape architect, Thomas Church produced groundbreaking design represented through fine arts photography in ways no one had ever seen: a synthesis of historical proportions.

The curatorial world took stock of the rise of landscape architecture as an art form. Between 1937 and 1957, the San Francisco Museum of Art, under the governance of the visionary director Grace L. McCann Morley, took the lead nationwide and organized four major exhibits devoted entirely to the genre. The first—only the third show that this institution put together since its birth in 1935—was titled *Contemporary Landscape Architecture and Its Sources*, and it came with a catalog gathering perspectives from notable figures such as Henry-Russell Hitchcock, Richard Neutra, and Fletcher Steele. Even with differing perspectives, the presence of the landscape in the making of architecture was collectively accounted for. From November 18 to December 26, 1948, a second exhibition, *Landscape Design*, provided an assessment of the developments in the field over a ten-year period, with an accompanying catalog filled with essays from different authors than in the first edition and notable built results. Here William Wurster, Thomas Church, Garrett Eckbo, Christopher Tunnard, and others refined the rhetorical register narrating the union between landscape and architecture. The third exhibition, *Landscape Architecture Today*, was held from August 16 to September 9, 1956, and it further developed the theme with additional designs. The final exhibition topping this twenty-year period was inaugurated the following year with the self-explanatory title *Landscape Architecture*. Conversely, it was not until 1964 that

the Museum of Modern Art in New York sponsored a critically acclaimed publication, *Modern Gardens and the Landscape*, written by Elizabeth Bauer Mock, sister of the public housing advocate Catherine Bauer (wife of architect William Wurster).

The climatic generosity of California might have played out in the attention given to landscape on the West Coast compared to the East in museum circles, but the substance of the matter is in no way conditional to a more forgiving climate. At any rate, the relationship between architecture and landscape shaped to a large extent the development of California Modernism, among others. The spatial consistency of the work of William Krisel in partnership with Dan Palmer in the Alexander Homes of Palm Springs, for example, can be explained by the fact that Krisel was licensed both as an architect and as a landscape architect. The fusion of architecture and landscape in the concurrent designs there was guaranteed from the onset, since the author of both was the same person.

Much has been said about the indoor-outdoor connection that Modern architecture opened up. As commonly shared, the directional nature of the connection, that is, from the inside to the outside, reveals the built-in assumption that architecture dominates the landscape. Landscape architects gradually recognized that the reverse was just as viable: outdoor-indoor, that is, from the outside to the inside, thus neutralizing this established hierarchy of architecture over landscape. In the goal of integration, the garden became an open framework to meet ever-changing functional needs, just as architecture was attending to. Garrett Eckbo further expanded these insights: "The landscape must be designed in toto, area by area, precisely because its quality is a direct result of the total combination of all elements seen from a given point of view or circulation pattern."[10]

With the Gould Garden in Berkeley, California, in the late 1950s, Lawrence Halprin's vision for the outdoor setting radically

OPPOSITE:
Stanton House, Belvedere, CA, 1961
Architect: Marquis & Stoller
Landscape Architect: Lawrence Halprin
Photographer: Ezra Stoller

ABOVE:
Adelman Residence, Los Angeles, CA, 1955–58
Architect: Thornton Abell
Landscape Architect: Bettler Baldwin
Photographer: Julius Shulman

THIS PAGE:
Cytron House, Beverly Hills, CA, 1964
Architect: Richard Neutra
Landscape Architect: Richard Neutra
Photographer: Julius Shulman

OPPOSITE:
Lehman House, Kent Woodlands, CA, 1961
Architect: Joseph Esherick
Landscape Architect: Lawrence Halprin
Photographer: Morley Baer

Edwards House, Beverly Hills, CA, 1957
Architect: Buff, Straub & Hensman
Landscape Architects: Wemple & Leone with Buff, Straub & Hensman
Photographer: Julius Shulman

transformed the bland character of the residence, which the owner had built a few years prior. As a technical challenge, due to the necessary use of multiple retaining walls, the dramatic drop in height from the entry is solved through a refined cascade of platforms in redwood and concrete, landing on a concrete tray with a six-sided pool inlaid in its boundaries. The circulation was carefully calibrated to deliver the sense of inhabiting an outdoor sculpture where the blend of hardscape and landscape yields a stage set from which to absorb the surrounding vistas. The strong architectonic nature of Halprin's design puts in crisis the disciplinary divisions of architecture and landscape. The qualitative nature of this open void far overrides the architectural merits of the preexisting structure, demonstrating the latent power of landscape architecture to radically sublimate given conditions into a coherent spatial statement of enduring strength. Such containment of space in the out-of-doors further reinforced the viability of extruding elements of the preexisting structure into a landscape reality existing as a stand-alone experience. The fountain and the scoring motifs on the retaining wall are of sculptor Jacques Overhoff's conception, whereas the pool pavilion is a collaboration with the architect owner, making this a veritable choral work under a landscape master plan.

While examples such as the Gould Garden increasingly made their appearance in the design vocabulary of the landscape community, architects, for their part, continued to reach out into the garden in their spatial ambitions. Austrian émigré Richard Neutra aimed at grounding his architecture to the specific site in virtually all his residential works. In a seminal article, he admitted that "the problem may have been narrowed down too much and the structure unjustly segregated from the total impact that it would produce when anchored to its surroundings."[11] He continued: "For ages buildings have been designed to exclude the elements to repel the atmospheric influences rather than to absorb them, as organisms do, for the vital process of assimilation and nourishment. . . . While manifestly a foreign body in the landscape, a building can nevertheless be virtually fused with it."[12] Neutra arrived at these conclusions as a mature architect, who had personally witnessed the rapid transitions that Modern architecture was undergoing. His realizations, however, were hardly common in earlier years or even at the time of that article.

In the relentless quest for a new image for architecture, the challenge for the modernists was how to negotiate the fixed image of a reformed architectural space with the changing image of the space of the landscape. The organization of the hardscape took the upper hand in the earlier period through geometric intricacies as a form of control to mitigate the unpredictable fluid language of plant materials and the natural flow of walking. Traditional landscape plans frequently offset the geometry of the house into the garden to replicate spatial relationships on the ground, often at a grander scale. A growing sense

Caygill House, Orinda, CA, 1950–51
Architect: George Rockrise
Landscape Architect: Lawrence Halprin
Photographer: Rondal Partridge

of the necessity to set up the environment for landscape to see architecture and for architecture to see landscape progressively took hold among architects. The transparency of Modern architecture called for a far closer relationship to landscape than ever before, its being a see-through invitation to let the environment in. If early Modern architecture had cutouts on its purist walls, then as decades passed and technologies developed, windows became walls of glass, obliterating that physical distinction between outside and inside for good.

The transfer of architecture concerns to the landscape from landscape concerns to architecture is partially seamless, however. Mies van der Rohe was seeking a common language in architecture. But can this thought be entertained in landscape architecture? Even the known distinction of servant and served spaces, as Louis Kahn articulated, appears tenuous in the outdoor setting. There are no real service spaces in landscape architecture. Space making in the outdoors is a likely candidate for expanded or altogether different rules of relation and aggregation than architecture. When, through Modern architecture, the new wave of landscape architects restructured the organization of the garden as a consequence of the internal conception of space, the garden was a place to experience rather than to look at, and therefore each area had to be designed for actual occupation. This is true even when function scripted the Modern garden, although it was hardly form determinant. Here the slogan "Form Follows Function" falls flat in the face of self-evidence. Using plant material to make space rather than embody predetermined geometry kept the field and its members anchored to a hybrid culture of historicism and modernity.

The Danish American landscape architect Jens Jensen, a contemporary of Frank Lloyd Wright, with whom he occasionally collaborated, articulated the spatial nuances of working with plant material for the definition of space from a different point of view than an arborist would take on. The space of landscape or landscape as space, with its own attributes independent of human intervention, constituted foreign notions in the copious literature on landscape architecture in Jensen's time. From his perspective, working with living materials gave the landscape architect a privileged position over the architect. Jensen wrote: "The study of curves is the study of life itself. Curves represent the unchained mind full of mystery and beauty. Straight lines belong to the militant thought."[13] He also wrote: "Landscaping is a composition of life that unfolds a mysterious beauty from time to time until mature age. All other arts are founded on dead materials."[14] While it is common to hear about design expression in architecture, it is rarer to encounter the notion of the expression of the land, positing the simultaneous presence of an architectural and land expression in the making of space.

How to make everything into one form is an opportunity of the fusion of landscape and architecture, currently negated through cultural and professional strictures. Garrett Eckbo—who stated in his oral history that when he attended the Landscape Architecture program at UC Berkeley in the early 1930s, he did not hear about Frank Lloyd Wright from the faculty—blamed this separation on a number of factors. Among them he identified the different licensing laws that architects and landscape architects have, along with different ways of working and cultural references, leaving the relationship between these two figures as complex as it was a century ago with the birth of modernity.

ABOVE:
Eckbo House, Los Angeles, CA 1959
Architect: Unknown
Landscape Architect: Garrett Eckbo
Photographer: Julius Shulman

OPPOSITE:
Vagtborg House, Los Angeles, CA 1951
Architect: Kenneth Lind
Landscape Architect: Unknown
Photographer: Julius Shulman

ABOVE & OPPOSITE:
Istanbul Hilton, Istanbul, Turkey, 1955
Architect: Skidmore, Owings & Merrill
Landscape Architect: Unknown
Photographer: Ezra Stoller

Footnotes

1. Frank Lloyd Wright, *The Living City* (New York: Horizon Press, 1958), 112.
2. Louis Kahn, *Drawings for City/2 Exhibition: Architecture Comes from the Making of a Room*, 1971.
3. Ludwig Mies van der Rohe, "I Don't Want to Be Interesting, I Want to Be Good," interview by John Peter, 1955, transcript, Pidgeon Digital, https://www.pidgeondigital.com/talks/i-don-t-want-to-be-interesting-i-want-to-be-good/.
4. Ibid.
5. Ibid.
6. Walter Gropius, "The Victory of the Modern Approach Is Sure," interview by John Peter, 1955, transcript, Pidgeon Digital, https://www.pidgeondigital.com/talks/the-victory-of-the-modern-approach-is-sure/.
7. James C. Rose, "Integration: Design Expresses the Continuity of Living," *Pencil Points* 19, no. 12 (December 1938): 759.
8. James Rose, *Gardens Make Me Laugh* (Norwalk, CT: Silvermine Publishers, 1965), 10.
9. William Wilson Wurster, "The Unity of Architecture and Landscape Architecture," in *Landscape Design* (San Francisco: San Francisco Museum of Art and Association of Landscape Architects, San Francisco Region, 1948), 6.
10. Garrett Eckbo, "Landscape Continuity," *Image* (student publication, School of Architecture, University of Texas) 3, no. 1 (May 1, 1965): 35.
11. Richard J. Neutra, "The Significance of the Natural Setting," *Magazine of Art* 43, no. 1 (January 1950): 18.
12. Ibid., 18.
13. Jen Jensen, *Siftings* (Baltimore: John Hopkins University Press, 1939), 34.
14. Ibid., 3.

ABOVE:
Graeme House, Los Angeles, CA, 1959
Architect: Buff, Straub and Hensman
Landscape Architect: Eckbo, Dean and Williams
Photographer: Julius Shulman

OPPOSITE:
Davis House, Midland, TX, 1966
Architect: Frank Welch
Landscape Architect: Unknown
Photographer: Ezra Stoller

PART 4

Landscape Architecture in Architectural History: The Grand Absent

In all times and places building and gardening have gone hand in hand.[1]

JOHN HARVEY

FOR CENTURIES, architectural history has placed primary importance on the individual buildings, the architects that designed them, and their patrons. Unless these structures are located in major city centers accessible to the general public, their context—that is, the fact that they exist as part of a physical setting bigger than themselves—has been decisively less of an object of critical appraisal. This historical practice is so embedded in the structure of the critical argument that its self-evidence has receded into the background of the collective consciousness.

Across the many and often competing ideologies feeding the work of past and current architectural historians, agreement is universal on one point. In the coming of age of Modern architecture, buildings shed weight off their load-bearing structures, becoming visually and environmentally permeable amid their surroundings more than ever before. More glass and less opaque volumes made their appearance in increasing number in the city fabric and in the countryside alike as the twentieth century unfolded. By turning buildings into see-through artifacts, they indirectly heightened their visual and spatial dependence on the areas directly outside their enclosures. In view of this much larger role assigned obliquely to what was happening right beyond the building's perimeter, the landscape near and far mattered at the least in the visual realm of the occupants looking from the inside out much more than previously experienced. The unprecedented porosity of buildings to the outside world was a sheer side effect of new construction technologies that emerged from the worldwide adoption of reinforced concrete, larger and larger glass plates, and, in the economies that were producing it, steel.

On the whole, the professional architectural historian crafts the reading of a piece of architecture by relying heavily on photographs (or renderings) of the completed building that portray its exterior images—that is, looking at the structure from the outside in—as well as on floor plans, and on models when available. Pictures of the interiors of the same projects being discussed make their entry in their essays only infrequently, leaving the option to provide a more in-depth account of the artifact entirely to the discretion of the historian's judgment. This practice is common around the globe, among historians of any persuasion. The architectural expression of each design statement takes center stage in the overall assessment, depending on the varying degrees of prominence that any new building has in its nearby context. Conversely, it is rare to find in those writings any mention of what the users of the building being examined are looking at in that acclaimed scheme: the design character of the open spaces and how those spaces relate to the architecture they are connected to. Given the generous fenestration displayed in decades of avant-garde projects, a generalized design choice among the modernists irrespective of the exterior climatic

PREVIOUS SPREAD:
Isbell House, Location Unknown, 1962
Architect: Unknown
Landscape Architect: Unknown
Photographer: Ezra Stoller

OPPOSITE:
House Beautiful Magazine Pace-Setter House of 1960
Corbett House, Cincinnati, OH, 1959
Architect: John Hill
Landscape Architect: Unknown
Photographer: Ezra Stoller

OPPOSITE:
House Beautiful Magazine Pace-Setter House of 1960
Corbett House, Cincinnati, OH, 1959
Architect: John Hill
Landscape Architect: Unknown
Photographer: Ezra Stoller

ABOVE:
Saxton Pope House, Orinda, CA, 1948
Architect: William Wurster
Landscape Architect: Unknown
Photographer: Ezra Stoller

conditions, it is legitimate to ask: In the mind of the creatives, what are the occupants, once inside those utopian visions, supposed to be viewing when directing their gaze outward? Are these new vistas made available to the modern individual delivering the intended sight of a reformed environment under modernist tenets?

Designing platforms for panoramic views and raising the horizon of the occupants inside and outside a piece of architecture were then, and continue to be, an ongoing theme in the field in all corners of the world. This belief in the inherent merits of opening the building up to the environment assumes, among other advantages, that there is something worth looking at out there, through large expanses of glass, beyond the building itself. Seeking legitimacy for what became a global hallmark, transparency at all costs, practitioners frequently made generic claims in the early days about the benefits that letting natural light come through had on health, and about the gains that viewing the neighboring scenery made for the mental states of the occupants. In the period when the mantra was to design the total environment, the liminal space surrounding the enclosure and extending deeper into the site took on a whole new relevance in the qualitative nature of experiencing modern space. In principle, the commitment of the architectural community to reconfigure the entire anthropized world under modernist guidelines gave no privileged point of view to this newly shaped setting, thus enabling the unambiguous visual connection between parts of the whole. Every place mattered. For instance, the Miesian space is an environmental continuum without leftovers. However, the critical tools adopted for historical analysis set the historians of modernity on a path that compromised the integral reading of the modernist intervention on the environment. In patterning the visual appreciation of architecture starting from the outside, looking toward the building, the magnification of that design took on a disproportionate dimension in relation to the whole, of which that same building is a subset. This unidirectional visual approach to architecture for the purpose of design evaluation was deep-seated, historically accrued, and came progressively at a price, utterly undetected at the onset of this practice. To put it succinctly, historians looked up and looked through, while rarely looking down to scrutinize the ground, where much human action takes place.

In opening any architectural history book, one sees that the single project, the city, and its makers are the protagonists of the curated selection of relevant facts for the construction of any narrative. Typically, in those textbooks, any conversation on landscape as space coexisting with the reality of the building has a marginal role in the argumentation and is hardly a topic of analysis. Routine mentions of landscape projects from the twentieth-century repertoire are exceptions, such as Park Güell in Barcelona, by Antoni Gaudí; the Woodland Cemetery in Stockholm, by Erik Gunnar Asplund and Sigurd Lewerentz; and the Sea Ranch Master Plan along the Northern California coast, by Lawrence Halprin. More rarely, but still within the architectural inquiry, is a design included such as the Path to the Acropolis in Athens, by Dimitris Pikionis, completed in 1957. But even in that case, the reflections of the architect himself on that project declared the vital role of the landscape. He wrote: "We rejoice in the progress of our body across the uneven surface of the earth. And our spirit is gladdened by the endless interplay of the three dimensions that we encounter at every step."[2] Architectural conditions are declared to be environmentally based, instead of staying strictly within the purview of the building alone.

Even rarer is the occasional featuring of the plans of the garden areas in those authoritative texts, constructing the written architectural culture of our time and the basis for canon formation. More typically, in the many editions of Sigfried Giedion's

ABOVE:
City Federal Savings Bank, Location Unknown, 1973
Architect: Edward Durell Stone
Landscape Architect: Unknown
Photographer: Ezra Stoller

OPPOSITE:
Erving Residence, Santa Barbara, CA, 1952
Architect: Lutah Maria Riggs
Landscape Architect: Thomas Church
Photographer: Maynard Parker

Leisure House, Duarte, CA, 1955
Architects: Douglas Honnold and John Rex
Landscape Architect: Thomas D. Church
Photographer: Julius Shulman

Space, Time, and Architecture, the Swiss historian paid attention to greenery and parks in the large conversation about town planning for both historical and modernist schemes, while making no mention of the role of landscape architecture in the formation of the New Tradition, as he aptly named the Modern movement. In later years, in the two volumes titled *Modern Architecture* by Manfredo Tafuri and Francesco Dal Co, the Sommerfeld residence and a concrete house in Berlin, both designed in 1921, are the only projects in the entire publication showing the garden designs together with the architecture. The same is true of other influential histories, such as those by eminent figures like Kenneth Frampton, William Curtis, David Watkin, Leonardo Benevolo, and Bruno Zevi, to name a few. The list is actually quite long. The history of Modern architecture is organized around fixed points

Mirman House, Arcadia, CA, 1959
Architect: Buff, Straub and Hensman
Landscape Architect: Buff, Straub & Hensman
Photographer: Julius Shulman

that exclude landscape as an agent of modernity. For example, prototypical design, a major chapter in modernist discourse, is a foreign notion to Modern landscape architecture, as each site is unique in its contextual commitment to the specific area it belongs to. Its design resolution is tied to a mixture of programmatic requirements and available material assets of the land configured in reciprocity with the architecture. There is little room in the history of the generalities of architecture for such attention to the specificity of the land design, impossible to become a model once conditions change. Even in more focused histories of particular experiences, this pattern is present. For instance, in her 1962 book *Modern California Houses*, Esther McCoy presented a unified account of the legendary Case Study Houses program in California, under the patronage and leadership of John Entenza from 1945 on. For more than twenty-five

Buff House, Pasadena, CA, 1980
Architect: Buff & Hensman
Landscape Architect: Conrad Buff
Photographer: Julius Shulman

Hensman House, Los Angeles, CA, 1976
Architect: Buff & Hensman
Landscape Architect: Unknown
Photographer: Julius Shulman

years, that title was the official document chronicling the development of a paradigmatic period in Southern California modernism. In it, the landscape architects who worked on the individual homes are credited in six of the built residences only. Saliently absent is their mention both for the unbuilt as well as the other built residences and in the "Biographies" secdtion of that book. In the missionary ideology of that enterprise, landscape architects were indirectly deemed as irrelevant, due to their cursory mention in the text. Furthermore, what is absent is even a cursory treatment of the formal character of the open spaces, despite numerous site plans that exhibit intentional design interventions in the gardens and indoor patios. The subtext of these books is that in an urban landscape, architecture alone constitutes its own unit of analysis included in the larger conversation about the city, with the landscape just an accessory to the main design statement,

ABOVE & OPPOSITE:
Moore House, Arcadia, CA, 1955
Architect: Burton Schutt
Landscape Architect: Ralph W. Smith
Photographer: Julius Shulman

a mere appendix to architecture. In this undeclared but ubiquitous outlook, the architecture of the landscape occupies the blind spot of architectural history. It is de facto invisible both to the writer and to the reader when the focus is only architecture. Many generations of architects have been trained in these ideas and internalized them, thus shaping their current practices.

The insistence on keeping the institutional separation between architecture and landscape among design practitioners and scholars in the official memory of the discipline is indeed peculiar, despite recent decades of debate about the rising awareness of the necessity to develop a lasting integration of architecture and site. Academic publications and commercial titles alike maintain this historically determined division between

ABOVE:
Richard Litsey House, Location and Date Unknown
Architect: Unknown
Landscape Architect: Unknown
Photographer: Ezra Stoller

OPPOSITE:
Marin County Civic Center, San Rafael, CA, 1962
Architect: Frank Lloyd Wright
Landscape Architect: Frank Lloyd Wright
Photographer: Ezra Stoller

Howard House, Palos Verdes Estates, CA, 1955
Architect: Burton Schutt and Harold Levitt
Landscape Architect: Henry Suenaga
Photographer: Julius Shulman

two concomitant drivers of space making, handing down their manufactured split through generations. This perspective remains pervasive, survives today in architectural historiography, and is as strong as ever. Landscape is mainly treated as a background to architecture instead of as an active agent determining the spatial identity of the entire environment under design, in accompaniment with architecture. This hierarchical conception of the built environment, architecture on top of landscape, lacks any experiential basis in the very evidence of the reality of building on the land. Furthermore, unless in an urban infill, the site is statistically much larger than the architecture itself, weighing considerably as an existing factor. Broadly speaking, architectural history is an extended chronicle of extrusions in space loosely connected through cultural affiliations or reactions against preceding aesthetic paradigms, whereas landscape architectural history is a longitudinal account of the design management of the ground.

Wasserman House, Palm Springs, CA, 1962
Architect: Harold Levitt
Landscape Architect: Unknown
Photographer: Julius Shulman

There is a powerful conundrum taking place in this rhetorical arrangement of design facts. It is a known circumstance that the direct experience of space reveals more than photography, or any mediated representation for that matter, can realistically capture in framing that same very space. Much of critical import is left out from the still view. It is also a logistical reality that it is highly unlikely a historian will visit all the sites he or she writes about in articles and books. The reliance on the photographic medium is therefore substantive. Photography curates the reality it represents, becoming a dominant influence on the first impact of the reality of the building onto lay and specialized readership alike. As articulated in Part One of this book, the perceptual displacement of photography in the viewer is a decisive player in organizing content through images. The vast majority of the times we look at an architectural photograph, we are exposed to an inverse representation of the environment where the building

dominates its setting, while real-world experience teaches us that the reverse is actually true.

With the arrival of modernity, a fundamental transition from image to space occurred. The overpowering roles of the elevation, the frontage, the facade as the drivers in the design process, determining the merit of any proposed project, were eventually relinquished after their peak in the nineteenth century; instead, the third (depth) and fourth (time) dimensions took over, upsetting the previous formal order and creating a new one. This had a profound impact on architecture, engineering, and landscape. It unleashed their latent potential for three-dimensional development in space, setting them in motion far apart from the static expression of the traditional built environment: volumes replaced planes across the board. In parallel with the rise of modernity, the

ABOVE & OPPOSITE:
Walker House, Los Angeles, CA, 1957
Architect: Harold Levitt
Landscape Architect: Mrs. Walker (house owner)
Photographer: Julius Shulman

technology of photography was born. Early on, the realism of the photographic image was taken at face value as a direct translation of the outer reality, without any loss of information. But even when later reasoning reframed this one-to-one relationship between representation and object represented, photorealism kept looming large in the viewer of photographic images. In a curious turn of events, photography reversed the newly energized architectural culture imbued with the discovery of those third and fourth dimensions back into an image again, making it captive of new perceptual limits.

To the dynamic response of the environmental experience when lived in, with the global circulation of images, civilization swapped the photograph, seen as a faithful token of the reality it was representing, and reinstated the supremacy of the fixed

viewpoint. That photographic perspective was, and still is, a variable entirely dependent on the skill and talents of the photographer, making the setting vulnerable to inadequate representation and subsequent perception. That perception, once socialized and historicized, establishes a hierarchy of values shared by local and global communities identifying themselves with particular design traditions. In de-skilling the body to the alertness of space, the flat experience of the image has dominated the transmission of design content ever since. Certainly photography, a medium for the selection of slices of reality and the creation of a new reality through the illusion of photorealism, helped modernist architects to make a case for the fitness of their designs within the existing context. However, it is the fluidity of the medium in carrying information for the viewer to internalize that weighs the most in this

ABOVE & OPPOSITE:
Los Angeles County Fair, garden and pool
Architect: Unknown
Landscape Architect: Harrison and Harrison
Photographer: Julius Shulman

argument. Since modernism set a new course in all the branches of the design field, either a piece of architecture or a garden is unaccountable for historical purposes in the absence of a photograph. However, historically the photograph has privileged the building at the expense of the landscape. Hence the perceptual loss of the environment in the visual culture of architecture.

A parallel approach from the landscape side determines an equally partial account of the environment. Historians of landscape architecture, from their end, contribute to reinforcing the divide between architecture and landscape. How the landscape registers in architecture and vice versa is a topic amenable to concurrent appropriation from historians of both disciplines. The steps leading from the raised courtyard to the grounds in the Säynätsalo Town Hall in Finland, by Alvar Aalto, and their echo in

Zall Garden, Yuba City, CA, 1955
Architect: Unknown
Landscape Architect: Thomas D. Church
Photographer: Unknown

the treatment of the landscape in the Sea Ranch Condominium in California, by Moore Lyndon Turnbull & Whitaker, are ideal examples of the enmeshment of landscape and architecture. Although these two projects are part of the architectural canon, the perspective of the landscape historian is hard to come by on those designs. While landscape architecture as a discipline exists in the absence of any architecture, the urban park being an obvious case, its interface with architecture is frequent enough to call for a systematic treatment that remains largely absent in contemporary historical practice. The lack of cross-pollination between these two communities fosters the perception of landscape and architecture as parallel endeavors with little crossover between them. While, in the practice of both professions, the dialogue between the two is a matter of course, it is the lack of a bridge or

Parks House, Beverly Hills, CA, 1961
Architect: Rex Lotery
Landscape Architect: Unknown
Photographer: Julius Shulman

common language that might be at the basis of the widespread mutual indifference to the broader scale.

The prominent architectural historian and preservationist James Marston Fitch, who held distinguished teaching credentials from Columbia University and published numerous architectural books, coauthored with F. F. Rockwell a text in 1956 titled *Treasury of American Gardens*.[3] It was a survey across the nations of the wide range of treatments of gardens, running from traditional arrangements to modernist layouts. Among the featured projects of the latter group, the Moore House in Ojai and the Tremaine House in Montecito, both in California, cover the garden design, with little emphasis on the architecture, even though they are two of the most celebrate residential works of Richard Neutra. The garden of the personal residence of the famous interior designer

Singleton House, Los Angeles, CA, 1960
Architect: Richard Neutra
Landscape Architect: Richard Neutra
Photographer: Julius Shulman

Alexander Girard in Santa Fe, New Mexico, is presented in later pages with little mention of the architecture, from his own design. In presenting the garden as a statement independent from the architecture, the perspective that garden and house could each be conceived in the absence of the other is further reinforced.

The writings of David C. Streatfield, focusing on two centuries of American landscape architecture, treat marginally the relationship between the design of the open spaces and architecture, zooming into the cultural and historical circumstances that influence the resolution of landscaped areas.[4] In the same vein, but more in depth, the encyclopedic *Design on the Land*, by Norman T. Newton,[5] was a highly praised chronicle of achievements in landscape architecture over millennia, dealing in greater depth with town planning as one of the developments of the

Time Space Study House, Location Unknown, 1958
Architect: Ramey and Himes
Landscape Architect: Unknown
Photographer: Julius Shulman

discipline but presenting rare points of contact between landscape and architecture for critical examination. The more recent *Landscape Design: A Cultural and Architectural History*, by Elizabeth Barlow Rogers,[6] is an ambitious survey that brings architecture closer to landscape than previous publications and traces along a very stretched timeline the various roles that landscape architecture occupied in the broader discourse of the built environment. In this unified framework, a sparse group of projects that are common to both architecture and landscape are featured, yet it clearly leans toward a reading of the natural and urban setting from the landscape perspective.

The inherited custom of this practice is that in the assessment of the environment, the focus is either on the landscape or on the architecture—hardly ever on both. In this either-or

mode of narration, the core of the argument is one of the two, architecture or landscape, with the authors of narratives from either scholarly community refraining from a unified account of both. Each camp of historians downplays the paramount role its counterpart plays in generating the environmental whole. Each marks its own disciplinary center in examining projects, at the expense of seeing them as part of an environmental ensemble permanently present both in landscape and in architecture. While they coexist in the real-world experience, they are orphans of a collective narrative. The disciplinary divide separating landscape architectural historians from architectural historians has split the

ABOVE:
Turner House, Los Angeles, CA, 1959
Architect: Burnett C. Turner
Landscape Architect: Burnett C. Turner
Photographer: Julius Shulman

OPPOSITE:
Monte Verde Model House, Los Angeles, CA, 1984
Architect: Smith & Williams
Landscape Architect: Unknown
Photographer: Julius Shulman

ABOVE & OPPOSITE:
John Deere and Company Corporate Headquarters, Moline, IL, 1961
Architect: Eero Saarinen
Landscape Architects: Hideo Sasaki and Stu Dawson
Photographer: Ezra Stoller

accounting of the environment in landscape and architecture, to the detriment of both. It has institutionalized in design culture the artificial breakup between the outside and the inside.

Two pairs of projects exemplify this point. Would it be conceivable to articulate a critical reading of the Upjohn Corporation General Office Building in Kalamazoo, Michigan, completed in 1967, and the Weyerhaeuser Corporate Headquarters in Federal Way, Washington, finished in 1971, two landmarks designed by Skidmore, Owings & Merrill, without also considering the role that Sasaki, Walker and Associates played as landscape architects in determining the character of the open spaces and the numerous courtyards? The IBM Corporation's Thomas J. Watson Research Center in Yorktown Heights, New York, of 1961, and the John Deere and Company Corporate Headquarters in Moline, Illinois, completed that same year, are two celebrated projects by Eero Saarinen & Associates.

Cannon House, Palm Desert, CA, 1963
Architect: William F. Cody
Landscape Architect: John Vogley
Photographer: Julius Shulman

Nonetheless, how often are the names of Hideo Sasaki and Stu Dawson mentioned as the landscape architects of these two schemes in the project credits and in the collective imagination, despite the outsize role of the setting in both?

Instances such as these abound in the extended chapters of the history of the built environment.

Yet, as Modern architecture took hold among design practitioners, it gave rise to unlikely associations where a hybrid professional space was created. Architects and landscape architects commingled, seeking multiple points of contact. Over time, for example, landscape architect Lawrence Halprin either employed

Dodd House, Laguna Beach, CA, 1963
Architect: Richard H. Dodd
Landscape Architect: Unknown
Photographer: Julius Shulman

architects, among them Steven Holl in the mid-1970s, or was the prime designer recruiting architects, as in the Arts Park project for California's San Fernando Valley, an eighty-five-acre park where Frank Gehry designed all the buildings. For their part, architects developed increasingly closer ties with landscape architects, further rebalancing this lopsided outlook. Dan Kiley was particularly sought after in his lifetime, given his ease in working on a large scale. Peter Rolland nurtured close collaborations with Edward Larrabee Barnes and Eliot Noyes. Harriet Pattison had significant influence on the late work of Louis Kahn.

Federal Reserve Bank, Detroit, MI, 1953
Architect: Minoru Yamasaki
Landscape Architect: Unknown
Photographer: Julius Shulman

In Latin America, Roberto Burle Marx's meteoric ascent to fame set him apart from his North American and European colleagues. His appearance in architecture and landscape architecture books makes him a bridge figure, even though he identified himself as a landscape designer with knowledge of painting, sculpture, and architecture. Partly due to his mother's love for plants that he was exposed to growing up, and partly due to his family's transfer to Berlin in the late 1920s, where he visited numerous botanical gardens and was deeply impressed by the paintings of Klee, Picasso, and Matisse, Burle Marx elaborated a distinct design idiom that linked a vast repository of plants and trees native to Brazil, through imaginative associations, to the

Adler House, Miami, FL, 1952
Architect: Rufus Nims
Landscape Architect: Unknown
Photographer: Ezra Stoller

FOLLOWING SPREAD:
Palo Alto City Hall, Palo Alto, CA, 1967
Architect: Edward Durell Stone
Landscape Architect: Edward D. Stone & Associates (EDSA)
Photographer: Ezra Stoller

aesthetic trends of the artistic avant-garde of Europe. He brought the language of emotions to his conceptions of the garden, making him a cult figure in his lifetime due to the singularity of his artistic vision. His opportunity to apply his knack for plant associations in the making of a garden came from a neighbor, architect Lúcio Costa, when in 1932 he was asked to design his first garden for a house the architect had just designed. Costa also introduced him to Le Corbusier, who had a profound impact on Burle Marx, including his articulating matters of space. Burle Marx's long association with the Brazilian master Oscar Niemeyer cemented his presence in the annals of twentieth-century modern architecture and landscape architecture alike. Burle Marx's unbounded love for indigenous

OUT
IN

HOTEL
HOTEL
Title Insurance
and Trust Company

Rockefeller Japanese House, Pocantico Hill, NY, 1978
Architect: Junzo Yoshimura
Landscape Architect: Junzo Yoshimura
Photographer: Ezra Stoller

plants provided him with his entry point into the field and determined his unique personal itinerary in the design richness of intentionally organized flora with an artistic end intertwined with architecture. For him, each garden was a different problem: a garden for nuns had to differ from one for children as well as from one for blind people.[7] He demonstrated to the worldwide community of Modernist architects how to marry in a garden space the ecological identity of the land with conquests of early twentieth-century art.

While architectural and landscape historians worked within the self-imposed confines of their disciplinary domains, a more inclusive understanding of the relationship between the two was being forwarded within the community of the practicing landscape architects. With an extensive body of writings, Garrett Eckbo, a seasoned practitioner and esteemed academician by the time of the following statement, provided a far more inclusive dimension to what historians had assigned to landscape until then. He wrote:

> Defined as open space design, landscape architecture has been concerned with all land not covered by buildings or engineering structures. . . . The problems of open space design have been problems of designing relationships between buildings and the gardens or parks around them. Pure landscape design, in the sense of large gardens or parks independent of buildings, came late in history, for people who had lost touch with natural open space, in its best aspects.[8]

He later continued in the same essay:

> The landscape is everything we see or sense wherever we are, indoors and outdoors, day and night, throughout our lives. The continuity in space makes physical elements more important than property lines or political boundaries.[9]

The suggested scope of landscape architecture made all sorts of areas, whose common attribute is the connection of the ground to the sky, candidates for a novel design approach. This revolutionary understanding of the outreach of landscape intervention empowered the landscape architect to take on a wide range of design challenges, ranging from the micro garden to vast territories. In the 1930s, Eckbo, Rose, and Kiley, started reimagining residual spaces, rehabilitating the remnants of the development process so as to restore their spatial dignity as extensions of houses in those early projects. This type of

Justin Day Residence, Sausalito, CA, 1950
Architect: Unknown
Landscape Architect: Thomas D. Church
Photographer: Maynard Parker

outlook had no historical representation when, even in modernity, the void contained in architecture took on a dominant role against the space with no roof, which inevitably was assigned a secondary role in the living experience. In this sense, architectural history turned into a potent echo chamber to spread across time inverted values on the hierarchy of the land versus the architecture, simply by bypassing the centrality of the open spaces in the quality of living in the built environment.[10]

The Norwegian architectural historian and critic Christian Norberg-Schulz endeavored to re-establish a link between landscape and architecture through his seminal book *Genius Loci: Towards a Phenomenology of Architecture*, published in 1979. Although

still leaning toward architecture, it did assign a central role to the land in establishing the character of a place and in becoming a determining factor in the production of architectural form. It was an idea that circulated very heavily throughout the 1980s, since the disconnect between architecture and the land, and the ensuing discontent of the general public, became unsustainable to leave the practice of architecture as such. The Nordic sensibility toward the landscape has been a constant, even before the arrival of modernity in the northern latitudes. From the early phase of Modern architecture, Nordic architects included landscape design in their architectural designs. Gunnar Asplund of Sweden, Aarne Ervi of Finland, Sverre Fehn of Norway, Arne Jacobsen of Denmark, and many others consistently marked their ground floor plans with a clear definition of the landscape of their buildings. It was an integral part of their practice way before any concern about the harmful effects of industrialization on the modern world.

William Prices Garden, Kentfield, CA, 1955
Architect: Unknown
Landscape Architect: Eckbo, Royston & Williams
Photographer: Ernest Braun

Occidental Center, Los Angeles, CA, 1965
Architect: William Pereira
Landscape Architect: Unknown
Photographer: Julius Shulman

Two further comments venture into an explanation for the hierarchical ambiguity between landscape and architecture.

The first observation is economic in nature. The subordinate role of landscape to the building in the city has roots in the evaluation of properties as investment opportunities and as objects of taxation. Property taxes for residences, for example, are determined based on the value of the individual house measured against comparable properties in size and location, assigning virtually no worth to the landscape other than the value of the land that the structure is sitting on. Whether the plot is landscaped or not has no bearing in determining its tax value or its market value, giving it at best an amenity role in private sales of the real estate asset. The presence of a park may or may not affect that value, and it is largely vulnerable to the location of that green lung in relation to the city it is part of. It follows that the economics of land use and property value inevitably push the available financial

resources toward improvement of the inventory of what is built aboveground, with the leftover of monetary efforts shared for the design of the landscape. As this imbalance is an integral part of the culture of building, the balance of values tends to favor the building and the architects, at the expense of the comprehensive understanding of the environment as a whole, inclusive of both the natural and the human made.

The second observation is tied to the digital tools currenly employed in the design process. Since the introduction of computers in design practice, the isolation of architecture from its environment is the consistent starting point of the architects' design exploration. When starting a new file in any 3-D modeling

ABOVE:
Thormin House, Los Angeles, CA, 1953
Architect: Anthony Thormin
Landscape Architect: Garrett Eckbo
Photographer: Julius Shulman

OPPOSITE:
Horton House, Los Angeles, CA, 1960
Architect: Harold Levitt
Landscape Architect: Edward Huntsman-Trout
Photographer: Julius Shulman

program, space is given as an absolute, suspended from any commitment to any physical circumstances that could influence its later shape. The neutral background in one color, defined through settings, triggers an automatic response in organizing shapes that are at their genesis self-referential. The relational nature of architecture to its physical adjacencies loses ground in the face of ubiquitous computing technology. To further complicate the matter, the realistic representation of actual trees with the minutiae of foliage and their casting shadows, especially when multiplied to define larger land arrangements, takes such an enormous amount of computing power that approximations are necessary to make the process manageable. The intersection of defaults in the construction of virtual space, limits in representation due to available

ABOVE:
Lavenant House, Pasadena, CA, 1953
Architect: Smith & Williams
Landscape Architect: Unknown
Photographer: Julius Shulman

OPPOSITE:
Howard House, Palos Verdes, CA, 1955
Architects: Burton Schutt and Harold Levitt
Landscape Architect: Henry Suenaga
Photographer: Julius Shulman

technologies, compounded by the neglect that landscape has traditionally suffered from in architecture when forwarding a comprehensive approach to the design of the environment, is a potent factor, albeit only one, in the estrangement of contemporary production of architecture.

The Finnish master architect Eliel Saarinen revealed the relational nature of architecture in his writings. He pointed out that architects should always draw what their building was part of—that is, its adjacency—so as to understand how it would relate to what was next to it. In this sense, the continuity of architecture is a matter of nested scales, facets of one integrated experience, where landscape and architecture finally become one.

ABOVE:
Edwards House, Escondido, CA, 1962
Architect: Hester & Davis
Landscape Architect: Unknown
Photographer: Julius Shulman

OPPOSITE:
Buck House, Beverly Hills, CA, 1956
Architect: Alfred Wilkes
Landscape Architect: Unknown
Photographer: Julius Shulman

Pool, Pasadena, CA, 1955
Architect: Unknown
Landscape Architect: Unknown
Photographer: Julius Shulman

Thomson Residence, Los Angeles, CA, 1950
Architect: Richard Spencer
Landscape Architect: Richard Spencer
Photographer: Julius Shulman

Footnotes

1. John Harvey, preface to *Mediaeval Gardens* (London: B. T. Batsford, 1981), xv.
2. *Dimitris Pikionis, Architect, 1887–1968: A Sentimental Topography* (London: Architectural Association, 1989), 9.
3. James Marston Fitch and F. F. Rockwell, *Treasury of American Gardens* (New York : Harper & Brothers, 1956).
4. David C. Streatfield, "Modernist Gardens 'on the Edge of the World,'" in *Masters of American Garden Design III:* "The Modern Garden in Europe and the United States"; *Proceedings of the Garden Conservancy Symposium Held March 12, 1993*, at the *Paine Webber Building in New York, New York*, ed. Robin Karson (New York: Garden Conservancy, 1994), 43–55.
5. Norman T. Newton, *Design on the Land: The Development of Landscape Architecture* (Cambridge, MA: Belknap Press of Harvard University Press, 1971).
6. Elizabeth Barlow Rogers, *Landscape Design: A Cultural and Architectural History* (New York: Harry N. Abrams, 2001).
7. Roberto Burle Marx, "A Garden Is Like a Poem," lecture, 1982, transcript, Pidgeon Digital, https://www.pidgeondigital.com/talks/a-garden-is-like-a-poem/.
8. Garrett Eckbo, "Theory of Landscape Design," National Council of State Garden Clubs, Landscape Design Conference, Sacramento, California, May 22–23, 1967, 1.
9. Ibid., 4.
10. Christian Norberg-Schulz, *Genius Loci: Towards a Phenomenology of Architecture* (New York: Rizzoli International, 1979).

ABOVE:
Kramer Residence, Norco, CA, 1953
Architect: Richard Neutra
Landscape Architect: Richard Neutra
Photographer: Julius Shulman

OPPOSITE:
Howard House Apartments, Malibu, CA, 1949
Architect: Richard Neutra
Landscape Architect: Richard Neutra
Photographer: Julius Shulman

Selected Bibliography

BOOKS

Aguar, Charles E., and Berdeana Aguar. *Wrightscapes: Frank Lloyd Wright's Landscape Designs*. New York: McGraw-Hill, 2002.
Appleton, Jay. *The Experience of Landscape*. New York: John Wiley & Sons, 1975.
Ashihara, Yoshinobu. *The Aesthetic Townscape*. Cambridge, MA: MIT Press, 1983.
Beauty for America: Proceedings of the White House Conference on Natural Beauty. Washington, DC, May 24–25, 1965. Washington, DC: U.S. Government Printing Office, 1965.
Bardi, Pietro Maria. *The Tropical Gardens of Burle Marx*. New York, NY: Reinhold Publishing Corporation, 1964.
Bognar, Botond, and Balazs Bognar. *Kengo Kuma: The Portland Japanese Garden*. New York: Rizzoli International, 2019.
Church, Thomas D. *Gardens Are for People*. New York: Reinhold, 1955.
——. *Your Private World: A Study of Intimate Gardens*. San Francisco: Chronicle Books, 1969.
Constant, Caroline. *The Modern Architectural Landscape*. Minneapolis: University of Minnesota Press, 2012.
Cullen, Gordon. *Townscape*. London: Architectural Press, 1961.
David, Jacques, and Jan Woudstra. *Landscape Modernism Renounced: The Career of Christopher Tunnard (1910–1979)*. London: Routledge, 2009.
Eckbo, Garrett. *The Art of Home Landscaping*. New York: F. W. Dodge, 1956.
——. Foreword to *Building in the Garden: The Architecture of Joseph Allen Stein in India and California*, by Stephen White. New York: Oxford University Press, 1993.
——. *Landscape Architecture: The Profession in California, 1935–1940, and Telesis*. Interviews by Suzanne B. Riess. Oral history transcript. Introduction by Robert N. Royston. Berkeley: Regional Oral History Office, Bancroft Library, University of California, Berkeley, 1993. https://digitalassets.lib.berkeley.edu/rohoia/ucb/text/landscapearch00eckorich.pdf.
——. *Landscape for Living*. New York: F. W. Dodge, 1950.
——. *The Landscape We See*. New York: McGraw-Hill, 1969.
Engel, David H. *Japanese Gardens for Today*. Foreword by Richard Neutra. Rutland, VT: Charles E. Tuttle, 1959.
Farr, Judith, with Louise Carter. *The Gardens of Emily Dickinson*. Cambridge, MA: Harvard University Press, 2004.
Fitch, James Marston, and F. F. Rockwell. *Treasury of American Gardens*. New York: Harper & Brothers, 1956.
Forty Gardens All to the Scale of 1:2400: Drawn by Forty-Two Design Students & Presented with Histories and Descriptions. Raleigh: Student Publication of the School of Design, North Carolina State University, 1967.
Frampton, Kenneth. "In Search of the Modern Landscape." In *Denatured Visions: Landscape and Culture in the Twentieth Century*, edited by Stuart Wrede and William Howard Adams, 42–61. New York: Museum of Modern Art, 1991.
Halprin, Lawrence. *Cities*. New York: Reinhold, 1963.
——. *Process: Architecture, No. 4*. Tokyo: Process Architecture, 1978.
Harrison, Robert Pogue. *Gardens: An Essay on the Human Condition*. Chicago: University of Chicago Press, 2008.
Harvey, John. *Mediaeval Gardens*. London: B. T. Batsford, 1981.
Hoffmann, Jens, and Claudia J. Nahson. *Roberto Burle Marx: Brazilian Modernist*. New Haven, CT: Yale University Press, 2017.
Howard, Hugh. *Architects of an American Landscape: Henry Hobson Richardson, Frederick Law Olmsted, and the Reimagining of America's Public and Private Spaces*. Boston: Atlantic Monthly Press, 2022.
Hunt, John Dixon. *Gardens and the Picturesque: Studies in the History of Landscape Architecture*. Cambridge, MA: MIT Press, 1992.
Imbert, Dorothée. *The Modernist Garden in France*. New Haven: Yale University Press, 1993.
Isohauta, Teija. *Alvar Aalto and the Art of Landscape*. Abingdon, UK: Routledge, 2022.
Jackson, John Brinckerhoff. *The Necessity for Ruins and Other Topics*. Amherst: University of Massachusetts Press, 1980.
——. *A Sense of Place, a Sense of Time*. New Haven: Yale University Press, 1994.
Jensen, Jen. *Siftings*. Baltimore: John Hopkins University Press, 1939.
Johnson, Jory. *Modern Landscape Architecture: Redefining the Garden*. New York: Abbeville Press, 1991.
Karson, Robin, ed. *Masters of American Garden Design III: "The Modern Garden in Europe and the United States"; Proceedings of the Garden Conservancy Symposium Held March 12, 1993, at the Paine Webber Building in New York, New York*. New York: Garden Conservancy, 1994.
Kassler, Elizabeth B. *Modern Gardens and the Landscape*. New York: Museum of Modern Art, 1964. Rev. ed., 1984.
Leatherbarrow, David. *The Roots of Architectural Invention: Site, Enclosure, Materials*. Cambridge: Cambridge University Press, 1993.
Lowell, Waverly B., Carrie L. McDade, and Elizabeth D. Byrne, eds. *Landscape at Berkeley: The First 100 Years*. Berkeley: University of California, Berkeley, College of Environmental Design, 2013.
Lurçat, André. *Terrasses et Jardins*. Paris: Editions d'Art Charles Moreau, 1929.
Marx, Leo. *The Machine in the Garden: Technology and the Pastoral Ideal in America*. New York: Oxford University Press, 1964.
McHarg, Ian L. *Design with Nature*. Garden City, NY: Natural History Press, 1969.
——. *A Quest for Life: An Autobiography*. New York: John Wiley & Sons, 1996.
McHarg, Ian L., and Frederick R. Steiner, eds. *To Heal the Earth: Selected Writings of Ian L. McHarg*. Washington, DC: Island Press, 1998.
Moore, Charles W., William J. Mitchell, and William Turnbull Jr. *The Poetics of Gardens*. Cambridge, MA: MIT Press, 1993.
Mosser, Monique, and Georges Teyssot, eds. *The Architecture of Western Gardens: A Design History from the Renaissance to the Present Day*. Cambridge, MA: MIT Press, 1991.
Mozingo, Louise A. *Pastoral Capitalism: A History of Suburban Corporate Landscapes*. Cambridge, MA: MIT Press, 2014.
Neutra, Richard. *Richard Neutra on Building: Mystery and Realities of the Site*. Scarsdale, NY: Morgan & Morgan, 1951.
Newton. Norman T. *Design on the Land: The Development of Landscape Architecture*. Cambridge, MA: Belknap Press of Harvard University Press, 1971.
Norberg-Schulz, Christian. *Genius Loci: Towards a Phenomenology of Architecture*. New York: Rizzoli International, 1979.
Olin, Laurie. *Essays on Landscape*. Amherst, MA: Library of American Landscape History, 2021.
O'Malley, Therese, and Joachim Wolschke-Bulmahn, eds. *Modernism and Landscape Architecture, 1890–1940*. Washington, DC: National Gallery of Art, 2015.
Owings, Nathaniel Alexander. *The American Aesthetic*. New York: Harper & Row, 1969.
Page, Russell. *The Education of a Gardener*. New York: Random House, 1962. 2nd edition, 1983.
Pikionis, Dimitris. *Dimitris Pikionis, Architect, 1887–1968: A Sentimental Topography*. London: Architectural Association, 1989.
Rancière, Jacques. *The Time of the Landscape: On the Origins of the Aesthetic Revolution*. Cambridge: Polity, 2023.
Raxworthy, Julian. *Overgrown: Practices between Landscape Architecture and Gardening*. Cambridge, MA. MIT Press, 2018.
Rogers, Elizabeth Barlow. *Landscape Design: A Cultural and Architectural History*. New York: Harry N. Abrams, 2001.
Root, Ralph Rodney. *Contourscaping*. Chicago: Ralph Fletcher Seymour, 1941.
Rose, James C. *Creative Gardens*. New York: Reinhold, 1958.
——. *Gardens Make Me Laugh*. Norwalk, CT: Silvermine Publishers, 1965.
Royston, Robert. *Robert Royston Oral History Interview Transcript*. Interviews by Charles A. Birnbaum, with J. C. Miller. Cultural Landscape Foundation, 2007. https://www.tclf.org/sites/default/files/atoms/files/RobertRoystonOralHistoryTranscript_FINAL.pdf.
Saarinen, Eliel. *Search for Form: A Fundamental Approach to Art*. New York: Reinhold, 1948.
San Francisco Museum of Art. *Contemporary Landscape Architecture and Its Sources*. exh. cat. San Francisco: San Francisco Museum of Art, 1937.
San Francisco Museum of Art and Association of Landscape Architects, San Francisco Region. *Landscape Design*. exh. cat. San Francisco: San Francisco Museum of Art and Association of Landscape Architects, San Francisco Region, 1948.
San Francisco Museum of Modern Art. *Lawrence Halprin: Changing Places*. exh. cat. San Francisco: San Francisco Museum of Modern Art, 1986.
Saunders, William S., ed. *Daniel Urban Kiley: The Early Gardens*. New York: Princeton Architectural Press, 1999.
Schildt, Göran, ed. *Alvar Aalto in His Own Words*. New York: Rizzoli, 1997.
Shepheard, Peter. *Modern Gardens: Masterworks of International Modern Architecture*. New York: Frederick A. Praeger, 1954.
Simonds, John Ormsbee. *Landscape Architecture: The Shaping of Man's Natural Environment*. New York: McGraw-Hill, 1961.
Sparke, Penny. *Nature Inside: Plants and Flowers in the Modern Interior*. New Haven, CT: Yale University Press, 2021.
Stappmans, Viviane, Nina Steinmüller, and Carolina Maddè, eds. *Garden Futures: Designing with Nature*. Weil am Rhein, Ger.: Vitra Design Museum, 2023.
Stilgoe, John R. *What Is Landscape?* Cambridge, MA. MIT Press, 2015.
Streatfield, David C. "History of Landscape Architecture: 1840–1930." In *Proceedings of the Landscape Design Conference Held January 15–16–17, 1968, at the Sacramento Municipal Utility District in Sacramento, California*, 1–2. National Council of State Garden Clubs.
——. "History of American Landscape Architecture: 1930–Present." In *Proceedings of the Landscape Design Conference III held January 20–21–22, 1969, at the Sacramento Municipal Utility District in Sacramento, California*, 1–7. National Council of State Garden Clubs.
——. "Modernist Gardens 'on the Edge of the World.'" In *Masters of American Garden Design III: "The Modern Garden in Europe and the United States"; Proceedings of the Garden Conservancy Symposium Held March 12, 1993, at the Paine Webber Building in New York, New York*, edited by Robin Karson, 43–55. New York: Garden Conservancy, 1994.
Sunset magazine editors. *Landscape for Modern Living*. Menlo Park, CA: Lane, 1958.
Taylor, Patrick, ed. *The Oxford Companion to the Garden*. Oxford: Oxford University Press, 2006.
Tishler, William H., ed. *Jen Jensen: Writings Inspired by Nature*. Madison: Wisconsin Historical Society Press, 2012.
Tuan, Yi-Fu. *Space and Place: The Perspective of Experience*.

Minneapolis: University of Minnesota Press, 1979.
——. *Topophilia: A Study of Environmental Perception, Attitudes, and Values*. Englewood Cliffs, NJ: Prentice-Hall, 1974.
Tunnard, Christopher. *Gardens in the Modern Landscape*. Postscript by Joseph Hudnut. London: Architectural Press; New York: Charles Scribner's Sons, 1938. Rev. ed., 1948.
——. *A World with a View: An Inquiry into the Nature of Scenic Values*. New Haven: Yale University Press, 1978.
Tunnard, Christopher, and Boris Pushkarev. *Man-Made America: Chaos or Control?* New Haven: Yale University Press, 1963.
Walker, Peter, and Melanie Simo. *Invisible Gardens: The Search for Modernism in the American Landscape*. Cambridge, MA: MIT Press, 1994.
Waymark, Janet. *Modern Garden Design: Innovation since 1900*. London: Thames & Hudson, 2003.
Whitehead, Alfred North. *The Concept of Nature*. Ann Arbor: University of Michigan Press, 1957.
Wrede, Stuart, and William Howard Adams, eds. *Denatured Visions: Landscape and Culture in the Twentieth Century*. New York: Museum of Modern Art, 1991.
Wright, Frank Lloyd. *The Living City*. New York: Horizon Press, 1958.

ARTICLES

Appleton, Jay. "Landscape Evaluation: The Theoretical Vacuum." *Transactions of the Institute of British Geographers*, no. 66 (November 1975): 120–23.
——. "Nature as Honorary Art." *Environmental Values* 7, no. 3 (August 1998): 255–66.
——. "Prospects and Refuges Re-visited." *Landscape Journal* 3, no. 2 (Fall 1984): 91–103.
Balmori, Diana. "Perspectives: Redefining the Boundary, Defining the Modern." *Progressive Architecture* 72, no. 8 (August 1991): 94–95.
Berque, Augustin. "Beyond the Modern Landscape." *AA Files*, no. 25 (Summer 1993): 33–37.
Carlson, Allen. "The Aesthetics of Environmental Architecture and Landscape, Territory, and Terrain." *Building Material*, no. 13 (April 2005): 60–63.
——. "Appreciation and the Natural Environment." *Journal of Aesthetics and Art Criticism* 37, no. 3 (Spring 1979): 267–75.
——. "Contemporary Environmental Aesthetics and the Requirements of Environmentalism." In "Environmental Aesthetics." Special issue, *Environmental Values* 19, no. 3 (August 2010): 289–314.
——. "Environmental Aesthetics and the Dilemma of Aesthetic Education." *Journal of Aesthetic Education* 10, no. 2 (April 1976): 69–82.
——. "Formal Qualities in the Natural Environment." *Journal of Aesthetic Education* 13, no. 3 (July 1979): 99–114.
——. "Nature, Aesthetic Appreciation, and Knowledge." *Journal of Aesthetics and Art Criticism* 53, no. 4 (Autumn 1995): 393–400.
——. "Nature, Aesthetic Judgment, and Objectivity." *Journal of Aesthetics and Art Criticism* 40, no. 1 (Autumn 1981): 15–27.
——. "On Appreciating Agricultural Landscapes." *Journal of Aesthetics and Art Criticism* 43, no. 3 (Spring 1985): 301–12.
——. "The Relationship between Eastern Ecoaesthetics and Western Environmental Aesthetics." *Philosophy East and West* 67, no. 1 (January 2017): 117–39.
——. "The Requirements for an Adequate Aesthetics of Nature." In "Environmental Aesthetics and Ecological Restoration." Special issue, *Environmental Philosophy* 4, nos. 1–2, (Spring/Fall 2007): 1–14.
Eckbo, Garrett. "Architecture and the Landscape." *Arts & Architecture*, October 1964, 22–23, 37–39.
——. "Collaboration between the Arts." Paper presented as part of "Architecture for Enjoyment," panel discussion at the 88th Annual Convention of the American Institute of Architects, Los Angeles, May 16, 1956, 1–4.
——. "Design in the Landscape." *Canadian Architect* 4, no. 6 (June 1959): 52–55.
——. "Detail in the Landscape." *Canadian Architect* 4, no. 7 (July 1959): 35–39.
——. "Landscape Continuity." *Image* (student publication, School of Architecture, University of Texas) 3, no. 1 (May 1965).
——. "Landscape Design in the U.S.A. as Applied to the Private Garden in California." *Architectural Review* 105, no. 625 (January 1949): 25–32.
——. "Landscape Gardening I: The Small Lot." *Architectural Forum* 84, no. 2 (February 1946): 76–80.
——. "Outdoors and In: Gardens as Living Space." *Magazine of Art* 34, no. 8 (October 1941): 422–27.
——. "Profession of Design: Design and the Landscape in White We Live." *Rockville* 10, no. 4 (August 1956): 47–49.
——. "Sculpture and Landscape Design." *Magazine of Art* 31, no. 4 (April 1938): 202–8, 250.
——. "Small Gardens in the City: A Study of Their Design Possibilities." *Pencil Points* 18, no. 9 (September 1937): 573–86.
——. "Theory of Landscape Design." National Council of State Garden Clubs. Landscape Design Conference, Sacramento, California, May 22–23, 1967.
——. "What Do We Mean by Modern Landscape Architecture?" *Journal, Royal Architectural Institute of Canada* 27, no. 8 (August 1950): 268–71.
Eckbo, Garrett, Daniel U. Kiley, and James C. Rose. "Landscape Design in the Primeval Environment." *Architectural Record* 87, no. 2 (February 1940): 74–79.
——. "Landscape Design in the Rural Environment." *Architectural Record* 86, no. 2 (August 1939): 68–74.
——. "Landscape Design in the Urban Environment." *Architectural Record* 85, no. 5 (May 1939): 70–77.
Eggener, Keith. "Postwar Modernism in Mexico: Luis Barragán's Jardines del Pedregal and the International Discourse on Architecture and Place." *Journal of the Society of Architectural Historians* 58, no. 2 (June 1999): 122–45.
Hall, William. "California Garden Is Outdoor Room." *San Francisco Examiner*, January 27, 1952, 1–2.
Halprin, Lawrence. "Hill Garden: The Importance of 'Edge.'" *Landscape Architecture* 50, no. 2 (Winter 1959–60): 96–99.
Hirsch, Alison. "Lawrence Halprin: The Choreography of Private Gardens." Special issue, *Studies in the History of Gardens and Designed Landscapes* 27, no. 4 (October–December 2007): 258–70.
Hunt, John Dixon. "The British Garden and the Grand Tour." In "Symposium Papers 10: The Fashioning and Functioning of the British Country House." Special issue, *Studies in the History of Art* 25 (1989): 333–51.
Jensen, Jens. "Novelty Versus Nature." *Landscape Architecture* 15, no. 1 (October 1924): 44–45.
——. "Roadside Planting." *Landscape Architecture* 14, no. 3 (April 1924): 186–87.
——. "Variations in Colors of Flowers." Letters. *Science*, n.s., 64, no. 1648 (July 30, 1926): 120.
McHarg, Ian L. "The Courthouse Concept." *Architects' Year Book*, no. 8 (1957): 74–102.
——. "Ecology, for the Evolution of Planning and Design." *VIA: Ecology in Design* (student journal, Graduate School of Fine Arts, University of Pennsylvania) 1 (1968): 44–67.
Messenger, Pam-Anela. "Thomas D. Church: His Role in American Landscape Architecture." *Landscape Architecture* 67, no. 2 (March 1977): 128–39, 170–71.
Neutra, Richard J. "The Significance of the Natural Setting." *Magazine of Art* 43, no. 1 (January 1950): 18–22.
Newton, Norman T. "Modern Trends: What Are They?" *Landscape Architecture* 22, no. 4 (July 1932): 302–3.
Oberlander, Cornelia Hahn. "Landscaping the Single Family House." *Canadian Architect* 1, no. 6 (June 1956): 21–25.
Parsons, Glenn, and Allen Carlson. "New Formalism and the Aesthetic Appreciation of Nature." *Journal of Aesthetics and Art Criticism* 62, no. 4 (Autumn 2004): 363–76.
Rose, James C. "Articulate Form in Landscape Design: People and Materials Defeat Preconceived Pattern." *Pencil Points* 20, no. 2 (February 1939): 98–100.
——. "Freedom in the Garden: A Contemporary Approach in Landscape Design." *Pencil Points* 19, no. 10 (October 1938): 639–43.
——. "Integration: Design Expresses the Continuity of Living." *Pencil Points* 19, no. 12 (December 1938): 759–60.
——. "Landscape Models." *Pencil Points* 20, no. 7 (July 1939): 438–40.
——. "Plants Dictate Garden Forms: Each Has Place as Material in Landscape Design." *Pencil Points* 19, no. 11 (November 1938): 695–97.
——. "Why Not Try Science: Some Tecnics [*sic*] for Landscape Production." *Pencil Points* 20, no. 12 (December 1939): 777–79.
Schwartz, Martha. "The Landscapes of Neglect." *Progressive Architecture* 72, no. 8 (August 1991): 96.
Spirn, Anne Whiston. "Seeing and Making the Landscape Whole." *Progressive Architecture* 72, no. 8 (August 1991): 92–94.
Tunnard, Christopher. "The Case for the Common Garden." *Architectural Review* 84, no. 502 (September 1938): 109–16.
——. "The Functional Aspect of Garden Planning." *Architectural Review* 83, no. 497 (April 1938): 195–99.
——. "Garden and Landscape 3: Shrubs for Garden Decoration in Winter." *Architectural Review* 85, no. 508 (March 1939): 147–48.
——. "Landscape Architecture in America." *Architects' Year Book*, no. 1 (1945), 30–38.
——. "Landscape into Garden: Reason, Romanticism, and the Verdant Age." *Architectural Review* 82, no. 491 (October 1937): 143–46, 151.
——. "Modern Gardens for Modern Houses: Reflections on Current Trends in Landscape Design." *Landscape Architecture* 32, no. 2 (January 1942): 57–64.
——. "Science and Specialization: The Conventional Garden of Today." *Architectural Review* 83, no. 495 (February 1938): 85–87.
——. "2. Scene." In "Man Made America." Special issue, *Architectural Review* 108, no. 648 (December 1950): 345–59.
Walker, Peter. "Lawrence Halprin & Associates, 1954: A Brief Memoir." *Landscape Journal* 31, nos. 1/2 (2012): 29–32.
Wilson, Colin St. John. "The Natural Imagination: An Essay on the Experience of Architecture." *Architectural Review* 185, no. 1103 (January 1989): 64–70.
"Wright as a Landscape Architect." In "Frank Lloyd Wright: His Contribution to the Beauty of American Life." Special issue, *House Beautiful*, November 1955, 342–47.

Photo and Drawing Credits

Ezra Stoller © Ezra Stoller/Esto: Front Cover, 1 (half-title page), 9, 11, 12–13, 14, 15, 17, 22, 24–25, 27, 28–29, 36–37, 42, 44–45, 46, 48, 50, 51, 53, 54, 55, 56–57, 58, 59, 60–61, 63, 68, 70–71, 78, 79, 80, 81, 82, 84, 85, 86, 88–89, 90–91, 92, 96, 98, 99, 100, 102, 103, 105, 106, 112, 114, 115, 122–123, 127, 128–129, 132, 133, 134, 136, 137, 139, 140, 141, 145, 146, 147, 149, 151, 154–155, 156, 166, 167, 169, 171, 172, 174, 175, 176, 184, 185, 198, 199, 203, 204–205, 206

Ernest Braun/Photograph by Ernest Braun: 2 (title page), 107, 108, 109, 138, 208

Morley Baer © 2024 The Morley Baer Photography Trust, Santa Fe: 16, 76, 159, 223

Rondal Partridge © 2024 Rondal Partridge Archive: 18–19, 39, 47, 64, 66–67, 72, 73, 87, 97, 116, 121, 124, 130, 148, 162–163

Julius Shulman photography archive, 1935–2009 © J. Paul Getty Trust. Getty Research Institute, Los Angeles: 21, 26, 30, 31, 32, 33, 34, 35, 40, 41, 43, 49, 52, 69, 93, 110–111, 113, 125, 126, 131, 135, 142, 143, 144, 150, 152, 153, 157, 158, 160, 164, 165, 168, 178, 179, 180, 181, 182, 183, 186, 187, 188, 189, 190, 191, 193, 194, 195, 196, 197, 200, 201, 202, 209, 210, 211, 212, 213, 214, 215, 216, 217, 218, 219

Glenn M. Christiansen, Photography: 23

Thomas D. Church (1902–1978) Collection, Environmental Design Archives, University of California, Berkeley: 38 (drawing)

Philip Fein, photographer. Courtesy of Thorne family Trust: 75

Lawrence Halprin Collection, The Architectural Archives, University of Pennsylvania: 77 (drawings)

Robert C. Cleveland, Photography: 83

Roger Sturtevant, photographer. Douglas Baylis (1915–1971) & Maggie Baylis (1912–1997) Collection, Environmental Design Archives, University of California, Berkeley: 94, 95

Theodore Osmundson, photographer. Theodore Osmundson Collection, Environmental Design Archives, University of California, Berkeley: 104

Maynard L. Parker, photographer. Courtesy of The Huntington Library, San Marino, CA 91108: Back Cover, 101, 119, 177, 207

Photograph on p. 192. Thomas D. Church (1902–1978) Collection, Environmental Design Archives, University of California, Berkeley

PAGE 224:
Gould Garden, Berkeley, CA, 1955–60
Architects: Gordon Reeve Gould with William Gillis and David Leaf
Landscape Architect: Lawrence Halprin
Photographer: Morley Baer

Acknowledgments

My decades-long exposure to architectural photography and my commitment to visit sites as frequently as possible provided the genesis for the birth of this book. In the countless trips to often obscure destinations to see both known and under-the-radar landmarks of twentieth-century architecture, I have frequently found myself spellbound by the beauty of the landscapes preceding the buildings I was actively tracking down. In these recurring experiences, I got a sense that architecture was part of the larger story of the environment to which that same building I was seeking was inextricably linked. This obvious fact consistently escaped my awareness, prompting me to examine the process behind my missed realization.

The choral nature of authoring a book is my recurring reminder that publishing is a collective endeavor. I am deeply indebted to Erica Stoller for her unwavering support in reading the manuscript and offering her pointed insights rooted in the experience of dealing with visual communication in architecture publishing. I am beyond thankful to Raymond Neutra and Eric Haesloop, who read different chapters and supplied me with critical observations that shaped later revisions of the text for the better. The conversations I had with Peter Walker and Tito Patri on landscape architecture and modernity have been beyond enlightening. I owe gratitude to both of them for their time and attention in answering my questions. I am grateful to Brett MacFadden, my graphic designer companion in previous book ventures. His ability to flex his creative talents in the face of constant updates and visual changes, consistently delivering a coherent outcome, humbles me. The infinite patience of my editor, Douglas Curran, in graciously maneuvering the bumpy road of manuscript delivery is the direct result of believing in the paramount importance of the role of landscape in the quality of our lives. I thank him for that. My wife, Silvia, has been next to me all along in the gestation of this project. Her dedication nurtured me during the writing process.

I dedicate this book to my brother Gianfranco Serraino (1964–2020).